Perspectives in Law, Business and Innovation

Over the last three decades, interconnected processes of globalization and rapid technological change—particularly, the emergence of networked technologies—have profoundly disrupted traditional models of business organization. This economic transformation has created multiple new opportunities for the emergence of alternate business forms, and disruptive innovation has become one of the major driving forces in the contemporary economy. Moreover, in the context of globalization, the innovation space increasingly takes on a global character. The main stakeholders—innovators, entrepreneurs and investors—now have an unprecedented degree of mobility in pursuing economic opportunities wherever they arise. As such, frictionless movement of goods, workers, services, and capital is becoming the "new normal".

This new economic and social reality has created multiple regulatory challenges for policymakers as they struggle to come to terms with the rapid pace of these social and economic changes. Moreover, these challenges impact across multiple fields of both public and private law. Nevertheless, existing approaches within legal science often struggle to deal with innovation and its effects.

Paralleling this shift in the economy, we can, therefore, see a similar process of disruption occurring within contemporary academia, as traditional approaches and disciplinary boundaries—both within and between disciplines—are being re-configured. Conventional notions of legal science are becoming increasingly obsolete or, at least, there is a need to develop alternative perspectives on the various regulatory challenges that are currently being created by the new innovation-driven global economy.

The aim of this series is to provide a forum for the publication of cutting-edge research in the fields of innovation and the law from a Japanese and Asian perspective. The series will cut across the traditional sub-disciplines of legal studies but will be tied together by a focus on contemporary developments in an innovation-driven economy and will deepen our understanding of the various regulatory responses to these economic and social changes.

The series editor and editorial board carefully assess each book proposal and sample chapters in terms of their relevance to law, business, and innovative technological change. Each proposal is evaluated on the basis of its academic value and distinctive contribution to the fast-moving debate in these fields.

More information about this series at http://www.springer.com/series/15440

Marcelo Corrales · Mark Fenwick
Nikolaus Forgó
Editors

Robotics, AI and the Future of Law

Springer

Editors
Marcelo Corrales
Center for Innovation and Research
Universidad Politécnica y Artística del Paraguay (UPAP)
Asunción, Paraguay

Nikolaus Forgó
Department of Innovation and Digitalisation in Law
University of Vienna
Vienna, Austria

Mark Fenwick
Faculty of Law
Kyushu University
Fukuoka, Japan

ISSN 2520-1875 ISSN 2520-1883 (electronic)
Perspectives in Law, Business and Innovation
ISBN 978-981-13-2873-2 ISBN 978-981-13-2874-9 (eBook)
https://doi.org/10.1007/978-981-13-2874-9

Library of Congress Control Number: 2018957672

This Springer imprint is published by the registered company Springer Nature Singapore Pte Ltd.
The registered company address is: 152 Beach Road, #21-01/04 Gateway East, Singapore 189721, Singapore

Preface

This volume is part of the book series: *Perspectives in Law, Business and Innovation*. The aim of this series is to provide a forum for the publication of cutting-edge research in the fields of innovation and the law from a Japanese and Asian perspective. The series aims to cut across the traditional sub-disciplines of legal studies, but is tied together by a focus on deepening our understanding of the various regulatory responses to technological, economic, and social change.

This volume constitutes the result of a joint cooperative effort drawing on the extensive global network of three academic institutions: The Center of Innovation and Research, part of the Universidad Politécnica y Artística del Paraguay (Asunción, Paraguay); the Department of Innovation and Digitalization in Law, part of the University of Vienna (Vienna, Austria); and the Graduate School of Law, part of Kyushu University (Fukuoka, Japan). Contributors to this book—including business and legal scholars and practitioners from Europe, East Asia, and the Americas—attempt to provide some of the latest thinking and assessment of current regulations with regard to robotics and emerging AI technologies.

The main target audience of the book comprises two different groups. The first group belongs to the legal community—particularly, legal scholars, law students, and practitioners—in the field of technology law who are interested in an up-to-date legal analysis of current trends. The second group are experts in the fields of AI, cloud computing, and robotics—including, service and infrastructure providers, IT managers, chief executive officers (CEOs), chief information officers (CIOs), and software developers—who are interested in, and influenced by, some of the shortcomings and benefits of the current legal issues under scrutiny in this work.

The editors would like to thank the editor-in-chief of this book series, Prof. Toshiyuki Kono, for opening the doors to this project and for his support. The editors are also indebted to the authors and co-authors of each chapter for their hard work, patience, and cooperation throughout the whole process from the initial concept to the final manuscript. Finally, the editors are grateful to the Springer staff for their support and efforts in ensuring final publication.

Asunción, Paraguay	Marcelo Corrales
Fukuoka, Japan	Mark Fenwick
Vienna, Austria	Nikolaus Forgó

Contents

The Rise of Robotics & AI: Technological Advances & Normative Dilemmas 1
Ugo Pagallo, Marcelo Corrales, Mark Fenwick and Nikolaus Forgó

Do We Need New Legal Personhood in the Age of Robots and AI? 15
Robert van den Hoven van Genderen

The Peculiar Case of the Mushroom Picking Robot: Extra-contractual Liability in Robotics 57
Ioannis Revolidis and Alan Dahi

Business and Regulatory Responses to Artificial Intelligence: Dynamic Regulation, Innovation Ecosystems and the Strategic Management of Disruptive Technology 81
Mark Fenwick, Erik P. M. Vermeulen and Marcelo Corrales

The Rise and Regulation of Drones: Are We Embracing Minority Report or WALL-E? 105
Pam Storr and Christine Storr

Profiling and Automated Decision-Making: Legal Implications and Shortcomings 123
Stefanie Hänold

Artificial Intelligence and Collusion: A Literature Overview 155
Steven Van Uytsel

Taming Artificial Intelligence: "Bots," the GDPR and Regulatory Approaches 183
Sam Wrigley

I, Inhuman Lawyer: Developing Artificial Intelligence in the Legal Profession 209
Dena Dervanović

Index 235

Editors and Contributors

About the Editors

Marcelo Corrales is an attorney-at-law specializing in intellectual property, information technology, and corporate law. He is also a professor and currently works as director of innovation and research at the Universidad Politécnica y Artística del Paraguay (UPAP). He has a Doctor of Laws (LL.D.) degree from Kyushu University in Japan. He also holds a Master of Laws (LL.M.) in international economics and business law from Kyushu University, and an LL.M. in law and information technology and an LL.M. in European intellectual property law, both from the University of Stockholm in Sweden. His most recent publications include *New Technology, Big Data and the Law* (Springer, 2017). His past activities have included being a research associate with the Institute for Legal Informatics and IT Law at Leibniz Universität Hannover (Germany) from 2007 to 2018.

Mark Fenwick is a professor of international business law at the Faculty of Law, Kyushu University, Fukuoka, Japan. His primary research interests are in the fields of white-collar and corporate crime, and business regulation in a networked age. Recent publications include *New Technology, Big Data and the Law* (Springer, 2017) and *The Shifting Meaning of Legal Certainty in Comparative and Transnational Law* (Hart, 2017). He has a Master's degree and a Ph.D. degree from the Faculty of Law, University of Cambridge (Queens' College) and has been a visiting professor at Cambridge University, Chulalongkorn University, Duke University, University of Hong Kong, Shanghai University of Finance and Economics, the National University of Singapore, Tilburg University, and Vietnam National University.

Nikolaus Forgó studied law at University of Vienna and Université Paris II Pantheon-Assas (Mag. iur 1990, Dr. iur. 1997). From 2000 to 2017, he was a full professor of legal informatics and IT law at Leibniz Universität Hannover (Germany), and between 2013 and 2017, he also served as data protection officer and chief information officer of this university. He was appointed as full professor of law at the University of Vienna in October 2017 and has been head of the newly founded Department of Innovation and Digitalization of Law since then. He teaches and consults in all fields of IT law, legal informatics, civil law, and legal theory and has been responsible for more than 50 research projects, including more than 20 EU research projects.

Contributors

Marcelo Corrales Center for Innovation and Research, Universidad Politécnica y Artística del Paraguay (UPAP), Asunción, Paraguay

Alan Dahi Grand Cayman, Cayman Islands

Dena Dervanović Stockholm, Sweden

Mark Fenwick Graduate School of Law, Kyushu University, Fukuoka, Japan

Nikolaus Forgó Department of Innovation and Digitalization in Law, University of Vienna, Vienna, Austria

Stefanie Hänold Institute for Legal Informatics, Leibniz Universität Hannover, Hannover, Germany

Ugo Pagallo Giurisprudenza, Università di Torino, Turin, Italy

Ioannis Revolidis Institute for Legal Informatics, Leibniz Universität Hannover, Hannover, Germany

Christine Storr Faculty of Law, Stockholm University, Stockholm, Sweden

Pam Storr Legal Consultant and Teacher in IT law, Stockholm, Sweden

Robert van den Hoven van Genderen Faculty of Law, Center for Law and Internet, Vrije Universiteit Amsterdam, Amsterdam, The Netherlands; Switch, Amsterdam, The Netherlands

Steven Van Uytsel Graduate School of Law, Kyushu University, Fukuoka, Japan

Erik P. M. Vermeulen Department of Business Law, Tilburg University, Tilburg, The Netherlands; Legal Department, Philips Lighting, Eindhoven, The Netherlands

Sam Wrigley Faculty of Law, University of Helsinki, Helsinki, Finland

Acronyms

AGI	Artificial General Intelligence
AI	Artificial Intelligence
ANI	Artificial Narrow Intelligence
ASI	Artificial Super Intelligence
ATS	Applicant Tracking System
BASIC	Beginner's All-purpose Symbolic Instruction Code
BMBF	Bundesministerium für Bildung und Forschung (Federal Ministry of Education and Research)
CEO	Chief Executive Officer
CIO	Chief Information Officer
CJEU	Court of Justice of the European Union
CV	Curriculum Vitae
DCFR	Draft Common Frame of Reference
DNS	Internet's Domain Name System
DPD	Data Protection Directive
DPIA	Data Protection Impact Assessment
EASA	European Aviation Safety Agency
EEG	Electroencephalogram
EGTL	European Group on Tort Law
ENIAC	Electronic Numerical Integrator and Computer
ESign	Electronic Signatures Act
EU	European Union
EuSEF	European Union Social Entrepreneurship Funds
EuVECA	European Union Venture Capital Fund
FCA	Financial Conduct Authority
FRII	Future Regulation of the Industrial Internet
FTA	Federal Trade Act
GDPR	General Data Protection Regulation
ICAO	International Civil Aviation Organization
IoT	Internet of Things

ISO	International Standardization Organization
M&A	Merger and Acquisition
MA	Model Aircraft
MS-DOS	Microsoft's Disk Operating System
NhRP	Nonhuman Rights Project
OECD	Organization for Economic Co-operation and Development
P2P	Peer-to-Peer
PETL	Principles of European Tort Law
PIA	Privacy Impact Assessment
ROA	Remotely Operated Aircraft
RPA	Remotely Piloted Aircraft
RPV	Remotely Piloted Vehicle
TCP	Transmission Control Protocol
TFEU	Treaty of the Functioning of the European Union
UA	Unmanned Aircraft
UAS	Unmanned Aerial Systems
UAV	Unmanned Aerial Vehicle
UCITA	Uniform Computer Information Transaction Act
UETA	Uniform Electronic Transaction Act
UK	United Kingdom
UN	United Nations
USA	United States of America
UUV	Unmanned Underwater Vehicle
VC	Venture Capital

The Rise of Robotics & AI: Technological Advances & Normative Dilemmas

Ugo Pagallo, Marcelo Corrales, Mark Fenwick and Nikolaus Forgó

Abstract Computer science, robotics and AI have all developed rapidly in recent years, bringing profound changes to all aspects of human life. However, the emergence and proliferation of these new technologies has not occurred within the bounds of traditional organizational, ethical and regulatory systems. We have reached an inflection point, where we need to pursue new business models and normative frameworks to underpin these fast-developing technologies. This introductory chapter briefly maps the evolution of these different technologies and argues for a new, more forward-oriented approach to the business and normative challenges that are created. The discussion ends with a review of the chapters that comprise this volume.

Keywords Machine learning · Robotics · Artificial intelligence
Normative challenges · Regulatory dilemmas

1 Introduction

The latest phase of the on-going "digital revolution" is characterized by self-reinforcing innovations in the fields of computer science, robotics, and AI. Technological developments in these areas have created complex new issues that need to be addressed by regulators and other policymakers. In the same way that cheap computational power and the increased availability of large amounts of data created

U. Pagallo (✉)
Giurisprudenza, Università di Torino, Turin, Italy
e-mail: ugo.pagallo@unito.it

M. Corrales
Center for Innovation and Research, Universidad Politécnica y Artística del Paraguay (UPAP), Asunción, Paraguay

M. Fenwick
Graduate School of Law, Kyushu University, Fukuoka, Japan

N. Forgó
Department of Innovation and Digitalization in Law, University of Vienna, Vienna, Austria

M. Corrales et al. (eds.), *Robotics, AI and the Future of Law*, Perspectives in Law, Business and Innovation, https://doi.org/10.1007/978-981-13-2874-9_1

ethical and regulatory challenges in previous decades, we are now experiencing a further wave of disruption. The interplay and synergies between the fields of computer science, robotics, and AI are particularly fruitful in this context in order to illustrate the importance of the business, ethical and regulatory challenges that are now emerging.[1]

To provide context for the chapters that follow, this chapter maps the evolution of different autonomous systems from the early stages of machine processing to more advanced robotics and AI, including virtual realities, sensors, algorithms, bots, drones, self-driving cars and more sophisticated "human-like" robots. It follows that we need a more nuanced and tailored approach to the business, ethical and regulatory challenges that are being created by these on-going technological changes, particularly if we wish to maximize the social and economic benefits of such technologies.

2 The Evolution of Computer Science and Machine Processing

In the field of computer sciences, one can, roughly and with simplifications, distinguish four different phases. The first period can be traced back to the work of the mathematician, Alan Turing. In his 1936 paper, *On Computable Numbers*,[2] Turing formalized a mathematical model of computation, according to which an abstract machine, namely a Turing machine, is theoretically capable of realizing all the tasks accomplishable by a computer, by manipulating symbols on a strip of tape in accordance with a set of pre-defined instructions. A decade later, in 1946, John von Neumann developed his hardware architecture for digital computers.[3] This occurred in the same year that Mauchly and Eckert presented their first electronic programmable computer, the ENIAC.[4]

Leaving the technical details aside, it is noteworthy that the Turing machine, the von Neumann architecture and the ENIAC, all hinged on the processing of numbers. (ENIAC means, after all, electronic numerical integrator and computer.) This number processing-approach is still at work in more recent developments in the field, such as Microsoft's operating system MS-DOS from 1982,[5] or the high-level programming language Altair BASIC that Paul Allen and Bill Gates developed in 1975.[6]

The second phase built on the first one, namely, developing from number processing to text processing. Two years after MS-DOS, in 1984, Microsoft released the text

[1] Pagallo (2013).

[2] Turing (1936).

[3] See, e.g., generally, Peragine (2013) and Wang (2008), p. 299.

[4] Eckert Jr., John Presper and Mauchly, John W.; Electronic Numerical Integrator and Computer, United States Patent Office, US Patent 3,120,606, filed 1947–06–26, issued 1964–02–04; invalidated 1973–10–19 after court ruling on Honeywell v. Sperry Rand.

[5] Saigh (1998), p. 162.

[6] Hey and Pápay (2015), p. 145, Kernighan (2017), p. 72.

editor Word for Macintosh. Five years later, in 1989, Tim Berners-Lee, working at the CERN labs in Genève, Switzerland, conceived a way to connect the technology of hypertexts to the Internet architecture and its protocols, such as the transmission control protocol (TCP), and the Internet's domain name system (DNS). By adding a new protocol, namely, the hypertext transfer protocol, or "http," between clients and servers of the information flow, a new network was born: the World Wide Web.

The third phase added to the previous ones another layer: from numbers and text processing to multimedia processing, such as the relational database MySQL from 1995 and the standards for audio/video codification and multimedia digitalization MP3 from 1997, and the MP4 a year later. (As early as 1999, the first in/famous peer-to-peer file sharing system on the Internet, Napster, was founded. By offering digital audio files, mainly songs encoded in MP3 format for free, however, Napster was found guilty of copyright infringement and forced to cease its operations in September 2002.)[7] Together with the diffusion of highly distributed P2P systems on the Internet, multimedia processing architectures also provided for services that seamlessly integrated text, sound, image, and video information, which are at the base of the current streaming industry.

The fourth phase concerns the convergence with robotics applications. Processing involves not only numbers, texts, or multimedia but also physical world processing, e.g., the 2008 robotic middleware provided by Willow Garage's ROS. By avoiding the shortcomings of traditional approaches, such as onboard computers for robots, the troubles with the computing power of such machines have increasingly been addressed by connecting them to a networked repository on the Internet, allowing robots to share the information required for object recognition, navigation and task completion in the real world.[8]

By closing the loop from robots to the Internet to robots that share information and learn from each other regarding their behavior and their environment, we need to identify how the field of robotics evolved, to converge with the science of computing. What were the parallel steps of this "other" discipline?

3 The Rise of Robots

In a similar way to our brief review of the evolution of computing processing, the development of robotics can be summed up with four different phases. First, robots emerged as reprogrammable machines operating in a semi- or fully autonomous way, to perform manufacturing operations. The first industry robot was tested within the automobile sector in 1961, drawing on the projects of George Devol and Joseph Engelberger, which culminated in the UNIMATE robot performing spot welding and extracting die-castings in a General Motors factory in New Jersey.[9]

---

[7]Rinsema (2017), p. 57; see also, generally, Chadwick (2006).

[8]See, e.g., Kharel et al. (2014), pp. 18–21.

[9]Pagallo (2013), Preface.

Then, the second phase occurred twenty years later, in the early 1980s, when the use of robotics within the car industry became critical. Japanese industry first began to implement this technology on a large scale in their factories, acquiring strategic competitiveness by decreasing costs and increasing the quality of their products. Western car producers learned a hard lesson and followed Japanese thinking, installing robots in their factories a few years later.

The third phase concerned the expansion of robotic applications for industry and professional services. As to industrial robots, such automatically controlled, reprogrammable, and multipurpose manipulator machines were increasingly employed in fields as diverse as refined petroleum products and nuclear fuel, textile and leather goods, communication and motor vehicles, agriculture and food products, and more. As to the professional service robots, they were progressively employed for inspection systems, construction and demolition, logistics, professional cleaning, defense, rescue and security applications, underwater systems, mobile platforms in general use, and so forth. And yet, even in the early 2000s, many individuals still had the impression that robotics was too dependent on the automobile industry. Remarkably, in the Editorial to the World 2005 Robotics Report of the Economic Commission for Europe and the International Federation of Robotics, Åke Madesäter raised this risk: "In the period 1997–2003, the automotive industry in Spain received 70% of all new robot installations. In France, the United Kingdom and Germany the corresponding figure amounted to 68, 64, and 57%, respectively."[10]

The fourth phase arose in the same years, as covered by the UN World report: the two-decade dependence of robotics on the automobile industry dramatically opened up to diversification, a "revolution" as described by many commentators.[11] This occurred with water-surface and unmanned underwater vehicles, (or "UUVs"), used for remote exploration work and the repairs of pipelines, oil rigs and so on, developing at an amazing pace since the mid-1990s. A decade later, unmanned aerial vehicles ("UAVs"), or systems ("UAS"), disrupted the military field.[12]

Thus, the final phase emerged in the 2010s, it was the turn of self-driving cars. Whereas the Governor of Nevada, in June 2011, signed a bill into law that for the first time authorizing the use of driverless cars on public roads, other states in the U.S. soon followed suit, up to the bill which the House of Representatives passed in September 2017, the Self Drive Act, that provides a much needed federal framework for the regulation of autonomous vehicles.

After the UUVs revolution, the UAS revolution, and that of self-driving cars, the range of robots now available suggests further candidates for the next robotic revolution in the field of service applications for personal and domestic use: robots for home security and surveillance, for handicap assistance, or for fun and entertainment.

In light of such diverse applications, we should not miss two crucial aspects of these technical developments. The first facet was already mentioned above in connection with the convergence between robotics and computer sciences: robots of

[10] UN (2005: ix).

[11] See, e.g., Gogarty and Hagger (2008) and Singer (2009).

[12] Pagallo (2011).

the fourth phase of robotics are increasingly connected to the Internet. Although this online connectivity makes a lot of sense to address the physical world-processing tasks of robots, it creates some new risks of its own, such as using these machines to perform malicious tasks under remote direction and/or to make them uncontrollable, e.g., via denial-of-service attacks.[13] After all, no sophisticated self-driving car was needed to make this threat clear in August 2015, when Fiat Chrysler had to recall more than a million vehicles after a pair of hackers showed they could take over a Jeep's digital systems via the Internet.[14]

The second crucial facet of this trend concerns the further convergence between robotics and a sub-field of computer science, that is, artificial intelligence ("AI"). Although spanning several disciplines, such as physics and mathematics, electronics and mechanics, neuroscience and biology, the field of robotics is increasingly intertwined with advancements of AI, to such an extent that even the definition of a "robot" has evolved over the fourth phase of the discipline. Some argue that we are dealing with machines built basically upon the mainstream "sense-think-act" paradigm of AI research.[15] Sebastian Thrun, former director of the AI Laboratory at Stanford, California, similarly reckons that robots are machines with the ability to "perceive something complex and make appropriate decisions."[16] While others stress that robots are those machines able to learn and adapt to changes in environments, it is worth discussing the steps that led to this convergence.

4 The Birth & Growth of AI

The birth of AI, namely, the design and setting of machines that mimic (also but not only) cognitive functions that humans associate with their own intelligence, such as learning and reasoning, planning and problem solving, is traditionally traced back to the workshop at Dartmouth in 1956 when, among others, Allen Newell, Herbert Simon, and J. C. Show presented their first AI program, i.e., the *Logic Theorist*. The expectations of both the founding fathers of the field and leaders of this kind of research were high. The aim was to create machines capable of doing all of the work that people can do, and to find the solution for the problem of attaining proper artificial intelligence, within "25 years"[17] or within "the current generation."[18]

However, AI had to pass through periods of stagnation, both financial and theoretical, that were later labeled as the "winter of AI." While the first occurred in the 1970s, a second winter also came between the late 1980s and early 90s, when the World Wide Web was thriving, and the first robotic revolution of UUVs was starting.

---

[13]Pagallo (2017).

[14]Grobman and Cerra (2016), p. 31.

[15]Bekey (2005).

[16]Singer (2009), p. 77.

[17]Simon (1965).

[18]Minski (1967).

In light of our previous remarks, the transformation of AI between the late 2000s and early 2010s comes as no surprise. The shift from simple automation to robust autonomous systems, in fact, partially overlaps with the final phases of computer sciences and robotics. Think about some sub-fields of AI, such as machine learning, namely, algorithms capable of defining or modifying decision-making rules autonomously, or the sector of data analytics, that is, the use of algorithms that make sense of huge streams of data.

Although we still do not have machines that are capable of doing all the work that people can do and the solution to the problem of creating proper artificial intelligence is not found yet, we are however increasingly dealing with systems that gain knowledge or skills from their own interactions with living beings inhabiting the surrounding environment, so that more complex cognitive structures emerge in the state-transition system of the AI application. Among the ingredients that made the convergence between computer sciences and robotics possible, we could list the improvement of more sophisticated statistical and probabilistic methods, the growing availability of vast amounts of data and massive computational power, up to the transformation of places and spaces into more IT-friendly environments, e.g., smart cities.

5 Mapping the Organizational, Ethical & Regulatory Dilemmas

These technological advancements have created enormous challenges, both for business and governments.

In an age of ever shorter innovation cycles, the next technological development is always looming. Firms already have to engage with the challenges created by robotics, automation and artificial intelligence. The pressures are incessant and anticipating the next "big thing" is crucial for maximizing a firm's chances of survival. Yet, meeting these challenges is much easier said than done. As companies grow, they tend to rely on hierarchical organizational structures. Such structures make sense as a strategy to manage the complexities of scale. The problem is that hierarchical organizations tend to result in a bureaucratic culture. This type of organization might have worked well in an era of mass-production, but is ill-suited to the business realities of today.[19]

A disconnect emerges between a ponderous, bureaucratic firm culture and the frictionless, dynamic character of a technology-driven economy. The inevitable effect of this disconnect is that "established" firms are unable to react effectively to the challenges created by fast-paced technological change. Such companies struggle to operate in markets where flexibility and speed are everything.

"Listed" companies, in particular, find it more and more difficult to keep pace. Regulatory pressures compound this problem. They create an unhealthy focus on

[19] See Fenwick and Vermeulen (2015).

maximizing shareholder value and short-term financial gains. The result is a more cautious, profit-driven culture. This may work during boom years or in stable markets. But today's interconnected global markets aren't stable and boom years are becoming less frequent as global markets create intense competition.

New technologies are similarly disruptive for government, particular existing ethical and regulatory frameworks. Again, this is not necessarily a new phenomenon. In the mid-1980s, for instance, the debate revolved around whether computers created new ways to control human actions, raising various ethical problems.[20] In the 1990s, lawmakers introduced the first provisions on computer crimes, data protection, digital copyright, and e-commerce. In the mid-2000s, the set of moral dilemmas arising from the evolution of robotics, and machine learning, suggested new domains of applied ethics, such as "robo-ethics,"[21] and "machine ethics."[22] As ethics is different from law, also the legal domain was challenged to give new, ethically convincing or at least justifiable answers. Accordingly, for example, EU lawmakers implemented a new general framework on data protection (in particular via the European General Data Protection Regulation, GDPR), specifically addressing topics such as automatic data processing, the right to an explanation of automated decision-making, and so forth. GDPR is only a part of a larger and prominently launched initiative to achieve a European Digital Single Market.

It seems reasonable to assume that the complexity of the normative problems created by new advancements in computer sciences, artificial intelligence, robotics, and their convergence, is only going to increase. This is, in large part, due to the synergies that are created as these technologies interact with one another.

6 Chapters

It seems clear, therefore, that AI, robotics and related technologies are disrupting the law and the legal profession. In particular, technological advances in fields such as machine learning, data mining, and automated reasoning are creating new and previously unimagined challenges for regulators, as well as new opportunities for legal professionals to make efficiency gains in the delivery of legal services. Given the exponential growth of such technologies, radical disruption seems likely to accelerate in the near future.

This collection therefore aims to bring together a series of contributions by leading scholars in the newly emerging fields of artificial intelligence, robotics and the law. The aim of this book is to enrich legal debates on the social meaning and impact of this type of technology.

Robert van den Hoven van Genderen opens the discussion by raising the fundamental question of robot identity and the law: Do we need to give robots and AI

[20] Johnson (1985) and Maner (1996).

[21] See, e.g., generally, Veruggio (2006).

[22] Wallach and Allen (2009).

entities a kind of legal personhood in a robotized society where activities with legal effect are increasingly performed by AI systems and autonomous robots? This question is considered by comparing the capacities and requirements of already existing legal subjects, natural persons and (artificial) legal persons such as corporations and states. The relevance of free will, intelligence and consciousness of natural persons to acquire legal personhood are analyzed and compared with other beings, animals and future AI entities.

Giving legal personhood to AI is also influenced by the human conviction that this would increase the risk to lose control and a "robot uprising." Man, as always, is afraid of technology getting out of hand and, therefore, wants retain control. In that context, the need for a certain legal personhood in a future legal framework, considering civil liability and even criminal liability is discussed.

Alan Dahi and *Ioannis Revolidis* focus on the issue of the extra-contractual liability of robots. The chapter suggests that robot-specific difficulties facing the legal system can be found in other areas of the law, and that the law has successfully addressed similar difficulties in the past. As such, a specific "Robot Law" is not necessarily needed. Moreover, the chapter argues that robots are too diverse a category to permit a uniform approach for dealing with the liability of their acts. Robots, and the underlying Artificial Intelligence, will need to be assessed against their specific purposes and capabilities.

The contribution does not intend to offer a detailed answer on how exactly the problem of extra-contractual liability of robots shall be addressed but instead offers a "first effort" to explore the methodological particularities of the problem.

As with the first chapter, this involves making analogies with earlier legal developments deemed to be of a similar character. The authors suggest that the lessons from regulating the Internet might point to a creative synthesis of technological advancements and traditional regulatory mechanisms, so that both are represented equally in the new set of rules that are meant to regulate new and disruptive phenomena, such as AI and robots.

The third chapter examines the interplay between business and regulatory responses to AI. As with other contributions, *Mark Fenwick*, *Erik P. M. Vermeulen* and *Marcelo Corrales* begin with the thought that identifying and then implementing an effective response to disruptive new AI technologies is enormously challenging for any business looking to integrate AI into their operations, as well as regulators looking to leverage AI-related innovation as a mechanism for achieving regional economic growth. These business and regulatory challenges are particularly significant given the broad reach of AI, as well as the multiple uncertainties surrounding such technologies and their future development and effects.

The chapter identifies two promising strategies for meeting this "AI challenge," focusing on the example of Fintech. First, "dynamic regulation," in the form of regulatory sandboxes and other regulatory approaches that aim to provide a space for responsible AI-related innovation. An empirical study provides preliminary evidence to suggest that jurisdictions that adopt a more "proactive" approach to Fintech regulation can attract greater investment.

The second strategy relates to so-called "innovation ecosystems." It is argued that such ecosystems are most effective when they afford opportunities for creative partnerships between well-established corporations and AI-focused startups and that this aspect of a successful innovation ecosystem is often overlooked in the existing discussion.

The chapter suggests that these two strategies are interconnected, in that greater investment is an important element in both fostering and signaling a well-functioning innovation ecosystem and that a well-functioning ecosystem will, in turn, attract more funding. The resulting synergies between these strategies can, therefore, provide a jurisdiction with a competitive edge in becoming a regional hub for AI-related activity. The chapter concludes with the suggestion that this approach can be relevant for the strategic management of other AI-related disruptive technologies.

In their chapter, *Pam Storr* and *Christine Storr* focus on a specific technology that has advanced significantly in recent years, namely drones. The popularity of drones has increased exponentially over the last few years. The advance of technology has not only led to drones being used by a greater number of actors in multiple settings, but has also allowed for the technological capacity of drones to increase at a great pace. Again, drones are interesting precisely because they involve the integration or gathering together of computer, robotic and AI technologies. As such, drones are an exemplary instance of the self-reinforcing character of modern technology described above.

How drone technology can be used, and by whom, has led to various regulatory dilemmas. The chapter then asks is the current legal framework equipped to cope satisfactorily with such technology and is it adapting with necessary changes at sufficient speed? The ways in which drones can, and have already, been used and misused have highlighted the need for certain regulatory developments in this field.

The chapter examines the various legal issues from a European perspective. It focuses on the diverse range of laws that are affected by the use of drones, specifically within the areas of surveillance, privacy and aviation. The underlying theme of the chapter focuses on whether the developing legal framework manages to deal successfully with the challenges surrounding the rapid rise in drone usage. The future of drones and the potential consequences of the legal framework being adopted are also addressed.

Stefanie Hänold's chapter also focuses on exploring the legal issues raised by a specific technology, namely automated decision-making in the context of "profiling." The increased use of profiling and automated decision-making systems similarly raises a number of challenges and concerns. The underlying algorithms embody a considerable potential for discrimination and unfair treatment. Furthermore, individuals are treated as "passive objects" of algorithmic evaluation and decision tools and are unable to present their values and positions. They are no longer perceived as individuals in their own right; all that matters is the group they are assigned to. Profiling and automated decision-making techniques also depend on the processing of personal data, and a significant number of the available applications are highly privacy-intrusive.

This chapter analyses how the GDPR responds to these challenges. In particular, Art. 22 GDPR, which provides the right not to be subject to automated individual decision-making, as well as the information obligations under Art. 13 (2) (f) and Art. 14 (2) (g) GDPR and the access right under Art. 15 (1) (h) GDPR, are examined in some detail. General data protection principles, particularly the principle of fairness, as well as specific German scoring provisions and anti-discrimination rules, are also looked at. In conclusion, various shortcomings of the present legal framework are identified and discussed and the outlook for possible future steps presented.

Steven van Uytsel's chapter is the third chapter to focus on a specific instantiation of an AI oriented technology, namely algorithms. The chapter focuses, in particular, on the competition law issues that such technologies raise, particularly in the context of collusion. The basis for this concern is the predicted change algorithms will bring to price setting. It is widely believed that algorithms, together with the gathering of Big Data, will increase the speed with which tacit collusion can be achieved and, in doing so, enlarge the market scope in which tacit collusion could be realized. The main question in this debate is whether competition law, as currently organized, can be applied to all scenarios in which algorithms have set the price.

The chapter offers a systematic review of the contemporary debate on this question. This debate received added impetus when Ezrachi and Stucke developed a new taxonomy to discuss algorithms and collusion. Their contribution on the topic triggered others to write on the issue, creating two different lines of thought.

On the one hand, there is the argument that technology is not yet sufficiently well-developed to let computers successfully collude without human intervention. That does not mean, however, that will not change in the future. On the other hand, there is a line of argument that algorithms will not necessarily evolve towards collusion. This part of the debate either suggests that algorithms may facilitate discriminatory pricing behavior or, in the worst case, result in other anti-competitive pricing strategies.

Presuming that algorithmic collusion can or may occur, a diverse set of solutions has been suggested. The most conservative one is to argue that the current law is broad enough to cover the technological evolution of algorithmic collusion. If this approach did not allow hard enforcement, other warning systems, sometimes backed up with fines, could be relied on. Others suggest developing a special rule of reason or a system to audit the algorithms. An alternative approach would be to enhance the privacy of consumers or to reduce price transparency, both with the aim of disabling systems to exploit their advantage in the market.

Whichever direction this issue evolves in the future, the literature suggests that this issue is developing in multiple directions. Even though the empirical evidence suggest that collusion is not currently likely, there is a broad consensus that artificial intelligence will progress. It is in preparation of such an event that the literature should develop possible ways of dealing with the technological progress.

Sam Wrigley focuses his discussion on a question that has received an enormous amount of public attention and is the focus of wideranging concerns on new technologies, namely the way that robots and AI have the potential to revolutionize the way that personal data is processed. Unlike processing performed by traditional methods, these new technologies have an unprecedented ability to gather, analyze and combine information.

The chapter describes how the introduction of "smarter" computers does not always mean that the actual nature of the processing is changed. Often, the result of machine-driven processing will be substantially similar to that performed by a human. We cannot, the chapter argues, regulate processing by robots and AI as a sui generis concept.

This chapter, therefore, examines the different regulatory approaches that exist under the EU's GDPR—the general regulatory approach (which treats all processing in the same way), the specific regulatory approach (which imposes specific rules for automated processing) and the co-regulatory approach (where data controllers are required to analyze and mitigate the risks on their own). It then considers how these approaches interact and makes some recommendations for how they should be interpreted and implemented in the future.

The final contribution, by *Dena Dervanović*, explores the important and related issue of how these fast-developing technologies will impact upon the legal profession. Legal technology—or "Legal Tech"—is changing the way lawyers practice law. In this context, Legal Tech refers to platforms, IT services, and software that first made law firms and lawyers more efficient in performing their activities. Practice management, document storage and automated billing and accounting software are prominent examples. Legal Tech also assists legal professionals in due diligence and discovery processes. Legal Tech has evolved from support systems to fully integrated and automated services for lawyers that increasingly disrupt the practice of law.

The chapter asks: what are the possibilities of having "AI lawyers" in the true sense—as autonomous, decision-making agents that can legally advise us or represent us? This discussion delves into the problematics and possibilities of creating such systems. This idea is inevitably faced with a multitude of challenges, among them the challenge of translating law into an algorithm being the most fundamental for the creation of an AI lawyer.

The chapter examines the linguistic aspects of such a translation and later moves to the ethical aspects of creating such lawyers and ethically codifying their conduct. This is followed by a discussion on whether Asimov's Three Laws of Robotics might be helpful in this regard. The ethical debate results in a proposal for a concept of Fairness by Design, conceived as the minimum standard for ethical behavior instilled in all AI agents. The chapter also attempts to give a general overview of the current state-of-the-art AI technologies employed in the legal domain as well as imagines the future of AI in Law. Subsequently, the chapter imagines an AI agent dealing with the "Solomon test" of splitting a baby. Finally, it is concluded that the advantage of having AI lawyers can be measured by the possibility of redefining the legal profession in its entirety as well as making legal advice and justice more accessible to all.

The distinctive feature of the contributions presented in this volume is that they address the impact of disruptive technological developments across a number of different fields of law and from the perspective of diverse jurisdictions. Moreover, the authors utilize insights from multiple related disciplines, in particular economics, social theory, and philosophy, in order to better understand and address the legal challenges created by robotics and AI. As such, the book highlights the inter-disciplinary character of debates on disruptive new AI technologies, robotics and their implications for the theory and practice of law.

References

Bekey, G. A. (2005). *Autonomous robots: From biological inspiration to implementation and control*. Cambridge, MA, London: The MIT Press.

Chadwick, A. (2006). *Internet politics: States, citizens, and new communication technologies*. Oxford: Oxford University Press.

Fenwick, M., & Vermeulen, E. P. M. (2015). The new firm: Staying relevant, unique & competitive. *European Business Organization Law Review, 16*, 595–623.

Gogarty, B., & Hagger, M. (2008). The laws of man over vehicle unmanned: The legal response to robotic revolution on sea, land and air. *Journal of Law, Information and Science, 19*, 73–145.

Grobman, S., & Cerra, A. (2016). *The second economy: The race for trust, treasure and time in the cybersecurity war*. New York: Apress.

Hey, T., & Pápay, G. (2015). *The computing universe: A journey through a revolution*. Cambridge: Cambridge University Press.

Johnson, D. (1985). *Computer ethics*. Englewood Cliffs, NJ: Prentice-Hall.

Kernighan, B. (2017). *Understanding the digital world: What you need to know about computers, the internet, privacy, and security*. Princeton: Princeton University Press.

Kharel, A. et al. (2014). Cloud robotics using ROS. *International Journal of Computer Applications*, 18–21. https://pdfs.semanticscholar.org/ccd1/ba285c69899c7fbf686679d0996a7d0bc25c.pdf. Accessed 23 Sept 2018.

Maner, W. (1996). Unique ethical problems in information technology. *Science and Engineering Ethics, 2*, 137–154.

Minski, M. (1967). *Computation: Finite and infinite machines*. Englewood Cliffs, NJ: Prentice-Hall.

Pagallo, U. (2011). Robots of just war: A legal perspective. *Philosophy and Technology, 24*(3), 307–323.

Pagallo, U. (2013). *The laws of robots: Crimes, contracts, and torts*. Dordrecht: Springer.

Pagallo, U. (2017). AI and bad robots: The criminology of automation. In M. R. McGuire & T. J. Holt (Eds.), *The Routledge handbook of technology, crime and justice* (pp. 643–653). London, New York.

Peragine, M. (2013). *The universal mind: The evolution of machine intelligence and human psychology*. San Diego: Xiphias Press.

Rinsema, R. (2017). *Listening in action: Teaching music in the digital age*. New York: Routledge.

Saigh, R. (1998). *The international dictionary of data communications*. Chicago: Grenlake Publishing Company Ltd.

Simon, H. (1965). *The shape of automation for men and management*. New York: Harper & Row.

Singer, P. (2009). *Wired for war: The robotics revolution and conflict in the 21st century*. London: Penguin.

Turing, A. M. (1936). On computable numbers, with an application to the entscheidungsproblem. *Proceedings of the London Mathematical Society, 42*(1), 230–265.

UN World Robotics. (2005). *Statistics, market analysis, forecasts, case studies and profitability of robot investment*. In UN Economic Commission for Europe (Ed.) and co-authored by the International Federation of Robotics. Geneva, Switzerland: UN Publication.

Veruggio, G. (2006). Euron roboethics roadmap. In *Proceedings Euron Roboethics Atelier*, February 27–March 3, 2006. Genoa, Italy.

Wallach, W., & Allen, C. (2009). *Moral machines: Teaching robots right from wrong*. New York: Oxford University Press.

Wang, Y. (2008). *Software engineering foundations: A software science perspective*. Boca Ratón: Auerbach Publications.

Do We Need New Legal Personhood in the Age of Robots and AI?

Robert van den Hoven van Genderen

Abstract Do we need to give robots and AI entities a kind of legal personhood in a robotized society where activities with legal effect are increasingly performed by AI systems and autonomous robots? In this chapter, this question is considered by comparing the requirements of existing legal subjects, natural persons and (artificial) legal persons such as corporations and states. The relevance of free will, intelligence and consciousness of natural persons to acquire legal personhood are analysed and compared with other beings, animals and future AI entities. To give legal personhood to AI is also influenced by the human conviction that this would increase the risk to lose control and a "robot uprising." Man, as always is afraid of technology getting out of hand and is convinced of their own superiority and therefore always wants to stay in control. In that context, the need for a certain legal personhood in a future legal framework, considering civil liability and even criminal liability is discussed as it is also subjected to considerations by the European Parliament, eventually leading to proposals in European law.

Keywords Artificial intelligence · Big Data · Ethics · Human control · Privacy GDPR · Legal personhood · Subject · Disruptive technologies · Liability Singularity · Robot law

---

This chapter is based on former articles, insights and presentations by the author and contributions by his students in the honors course "Robotlaw" of the combined VU and UvA Universities. The terms "robot" and "AI entity" are used interchangeably.

R. van den Hoven van Genderen (✉)
Faculty of Law, Center for Law and Internet,
Vrije Universiteit Amsterdam, Amsterdam, The Netherlands
e-mail: rob.vandenhovenvangenderen@switchlegal.nl

R. van den Hoven van Genderen
Switch, Singel 117, 1012 VH Amsterdam, The Netherlands

M. Corrales et al. (eds.), *Robotics, AI and the Future of Law*, Perspectives in Law, Business and Innovation, https://doi.org/10.1007/978-981-13-2874-9_2

1 Introduction

"I believe that the abominable deterioration of ethical standards stems primarily from the mechanization and depersonalization of our lives," "a disastrous byproduct of science and technology. Nostra culpa!".[1]

Where does this thought ultimately lead us? Will we as the unique human race disappear in "singularity"?[2] Sofia, the lifelike robot from Hanson Robotics received citizenship from Saudi Arabia in the autumn of 2017. Is this the one swallow for the AI summer? Should we fear or welcome such a development? What will be the future of mankind in a state of techno-predictive determinism? There is a concern that humanity will not be able to control AI or, at least, will not be able to predict the behavior of self-learning robots. Will the developers of AI and Robots be the "new Oppenheimers"? How do we control the development of AI? Is the embedding of AI and robotics in our legal system a viable solution or will AI create its own legal system; and will we be transformed into objects instead of subjects? Will the development of autonomous systems create "killer robots," as warned by several scholars and leaders of industry? Do we have to stop all further innovation of AI in order to save mankind? Should we seek to stop this technological evolution? Or, is it acceptable to integrate AI fully in our society and also in our legal system?

As always the case with new technological developments, there is a fear of the negative consequences of introducing AI and robotics throughout society. Loss of jobs, loss of control and ultimately fear for the future of mankind. We also have the tendency to accentuate the negative aspects of any new technology, certainly when we do not fully understand the technology and its consequences. As Sir Arthur Clarke stated in his novel Profiles of the Future: An Inquiry into the Limits of the Possible his so-called third law: any sufficiently advanced technology is indistinguishable from magic.[3]

And magic is incomprehensible and, therefore, dangerous: as Steven Hawking, Elon Musk and others have warned us: AI is the "biggest risk we face as a civilization" and "AI is a rare case where we need to be proactive in regulation instead of reactive because if we are reactive in AI regulation it's too late, AI is a fundamental risk to the existence of civilization…as a whole."[4]

Also, Vladimir Putin has emphasized how AI could be the subject for a new "arms race" in a speech for the opening of the school year in Russia: "Artificial intelligence is the future, not only for Russia, but for all humankind…It comes with

[1] Alfred Einstein in a letter he wrote to his friend, psychiatrist Otto Juliusburger, in 1948.

[2] The acceleration of technological progress has been the central feature of this century. We are on the edge of changes comparable to the rise of human life on Earth. The precise cause of this change is the imminent creation by technology of entities with greater [intellectual capacity] than human intelligence. See Vinge (1993).

[3] Clarke (1973).

[4] Titcomb (2017) AI is the biggest risk we face as a civilization, Elon Musk says. Available at: http://www.telegraph.co.uk/technology/2017/07/17/ai-biggest-risk-face-civilisation-elon-musk-says/. Accessed 11 October 2017.

colossal opportunities, but also threats that are difficult to predict. Whoever becomes the leader in this sphere will become the ruler of the world." Though, according to Putin, we do not have to be worried: "If we become leaders in this area, we will share this know-how with the entire world, the same way we share our nuclear technologies today."[5] This statement did not satisfy Elon Musk who "twittered" the following in response to the words of Putin: "As China, Russia, soon all countries with strong computer science. Competition for AI superiority at national level most likely will be the cause of WW3."

Not losing ourselves in war-scenarios, we have to come back to the question if and how we can regulate these technological developments on a national and international level. Until now, law has been developed by humans, for humans and—initially—to govern the relations between natural persons and, later on, artificial legal persons. But many things have changed during the historical development of the law, in the long journey from the Roman legal system to our modern legal system. New technologies will change society and will reflect on the change of this legal framework. As Lauren Burkhart citing Clark A. Miller and Ira Bennett on "reflexive governance," observes we better be prepared to have an open mind for changes in technology by "identifying not only what gadgets might arise but also how gadgets intersect in society, with one another and with people, how people identify with, make use of, oppose, reject, apply, transform, or ignore [technologies]."[6]

To what extent society must adapt to technological innovations has to be based on the needs of that society, be it economic or social. If a sentient entity, in the sense of possible autonomous intelligent agency in robotics and other AI systems, now, or in the near future, could be expected to act with legal effect, that is to say perform tasks with legal consequences, the legal framework could be adapted accordingly. This decision, however, should assume that an AI robotized society will benefit from—to a certain degree—the legal personality of robots. Legal scholars are generally hesitant to adapt the law on the basis of technological changes. But "if the facts too long deviate from the legal status and the right is unsustainable, the law must ultimately yield to the actual situation."[7]

Already society has undergone changes as a result of this development. Semi-autonomous cars are now a point of legal, moral and social discussion because the central subject in traffic laws is the driver and their control over the vehicle is a requirement for safety on the road. This gives rise to a question that is not new, nor solely legal; a question that was already described by Geldart in a discipline overruling way:

---

[5] Vincent (2017) Putin says the nation that leads in AI "will be the ruler of the world." Available at: https://www.theverge.com/2017/9/4/16251226/russia-ai-putin-rule-the-world. Accessed 11 October 2017.

[6] Lauren Burkhart citing Miller and Bennett (2008) and Burkhart (2016).

[7] Tjong Tjin Tai (2016), p. 248.

> The question is at bottom not one on which law and legal conceptions have the only or the final voice: it is one which law shares with other sciences: political science, ethics, psychology, and metaphysics.[8]

It is of the utmost importance to also consider ethical values and fundamental rights issues in the possible decision to give a certain legal status to robots. Neil Richards and Jonathan King's statement in their paper on Big Data ethics could well be applied to robotics:

> We are building a new digital society, and the values we build or fail to build into our new digital structures will define us. Critically, if we fail to balance the human values that we care about, like privacy, confidentiality, transparency, identity and free choice with the compelling uses of Big Data, our Big Data Society risks abandoning these values for the sake of innovation and expediency.[9]

2 Legal Subjects as Responsible Actors

Although natural persons and legal persons have, for a long time, been the key players in our legal system this has not always been the case. Large and small businesses, private organizations and government organizations are entitled to carry out all kinds of acts as legal entities and can be held responsible for the things they do. But, in the Middle Ages, for instance, animals could also be held responsible for their acts.[10] Technological development develops in the direction of artificially intelligent programs possibly embodied in all kind of physical instruments and a variety of robotic entities in more or less anthropomorphic shapes that can perform a variety of tasks. Coupled with the exponentially expanded Internet, decision-making by these AI entities with legal consequences is creeping up to us. The consideration whether an autonomously functioning artificial intelligent entity or robot must have a certain legal subjectivity or not, will be dependent upon social and economic necessities and not least of all, the cultural social and legal acceptance by other actors. In other words, can a future society function without any form of legal personality for autonomous, artificially intelligent entities or is it a "conditio sine qua non?"

It is important to consider what kind of reasoning will be applied to the determination of the legal status of AI and robots. This status could be built on an augmented layer of required legal elements based on the continuous development of autonomy and intelligence of the robot. Or one could analyze the characteristics of the current players with legal personality and select which elements will be desirable to give robots that degree of legal personality that is considered useful in society.

Cautious proposals are already being made to comply with the future and to find legal solutions. However, the actual legal implications of an AI integrated society are set aside. Although the European Parliament accepted a motion on the civil law

---

[8] Geldart (1911), p. 94.

[9] Richards and King (2014), p. 394.

[10] Berriat Saint-Prix (1829).

aspects of the development of AI generated robotics, in creating electronic legal personhood, it is at a rather high level of abstraction:

> 59 f) creating a specific legal status for robots in the long run, so that at least the most sophisticated autonomous robots could be established as having the status of electronic persons responsible for making good any damage they may cause, and possibly applying electronic personality to cases where robots make autonomous decisions or otherwise interact with third parties independently.[11]

The essence though is recognized: legal interaction with other parties. The orientation on electronic (legal) persons though is limiting the possibility of application of other future technologies.

3 What About AI and Robots

For an analysis of the legal positioning of robots and AI, we cannot escape defining or describing these phenomena. Of course, there are several definitions developed by scientists and lawyers. For the sake of clarity, this chapter will not delve into all of these conceptions. There are a range of robots varying from the simple one task-oriented industrial robot to the autonomous car and the anthropomorphic robot companion. Bertolini defined a robot in a broad sense, encompassing this wide variety of robotics and AI entities as follows:

> A machine, which (i) may be either provided of a physical body, allowing it to interact with the external world, or rather have an intangible nature – such as a software or program, – (ii) which in its functioning is alternatively directly controlled or simply supervised by a human being, or may even act autonomously in order to (iii) perform tasks, which present different degrees of complexity (repetitive or not) and may entail the adoption of not predetermined choices among possible alternatives, yet aimed at attaining a result or provide information for further judgment, as so determined by its user, creator or programmer, (iv) including but not limited to the modification of the external environment, and which in so doing may (v) interact and cooperate with humans in various forms and degrees.[12]

In determining the need for the legal personhood of AI entities, it should be considered that these systems will clearly vary in function. There will be obvious differences in

[11]Whereas it is of vital importance for the legislature to consider all legal implications. All the more now that humankind stands on the threshold of an era in which ever more sophisticated robots, bots, androids and other manifestations of AI seem poised to unleash a new industrial revolution that is likely to leave no stratum of society untouched; report Delvaux with recommendations to the Commission on Civil Law Rules on Robotics (2017); European Parliament resolution of 16 February 2017 with recommendations to the Commission on Civil Law Rules on Robotics (2015/2103(INL)) <A8-0005/2017>.

[12]Bertolini (2013), p. 219. Compare also the definition by "robotpark": "A robot is a mechanical or virtual artificial agent (called "Bot"), usually an electro-mechanical machine that is guided by a computer program or electronic circuitry. Robots can be autonomous, semi-autonomous or remotely controlled and range from humanoids such as ASIMO and TOPIO, to nanorobots, "swarm" robots and industrial robots. A robot may convey a sense of intelligence or thought of its own."

the degree of autonomy resulting in a variety of legal requirements dependent on a social need to have robots perform tasks as more or less autonomous acts.

For the possible legal analysis and classification of robots, it is required to look at: (1) the embodiment or nature of the robot; (2) the degree of autonomy; (3) the function of the robot; (4) the environment; (5) the nature of the interaction between human and robot.[13]

On the basis of these considerations, we can formulate the following questions:

Is there a need for a framework for AI and robot law in the sense of a law relating to, or as a result of, the use of robot technology in society? And, if so, what are the preconditions for establishing such a law in our legal system?
Does the robot need a certain degree of legal personhood that does not yet exist in positive law and is it necessary to regulate that degree of legal personhood? And, if so;
Is there a "gradation" of legal embodiment that connects with existing forms of legal personality or is a sui generis construction desirable considering the variability of AI systems and robotics?

4 The Essence of Legal Personhood

Before looking further into the question of what legal personhood would mean for an autonomous robot, one has to understand what it actually means to have legal personality, or in other words, what does it mean to be considered a legal person. The technical legal meaning of being a legal person is in a simplified version: "*a subject of legal rights and duties.*"[14]

This does not necessarily refer to "natural persons." The idea of legal personhood involves the status of an entity as a person before the law, leading to recognition of certain rights and obligations under the law. Consequently, a legal person has the duty to obey the law, while enjoying the benefit of protections to rights and privileges accorded to a legal person.

In Dutch law, for instance, no specific definition is given but it can be understood as being capable of having legal rights and duties and legal capacity within a legal system, to act with legal effect such as to enter into contracts, to be liable, to be a subject of legal remedies. A legal (artificial) person is considered equal to a natural person, as far as property law is concerned, unless the law explicitly states the contrary.[15]

The legal construct of personhood in the law, however, operates as a bundle of fundamental assumptions involving the biological understanding of human beings,

---

[13] Bertolini (2013).

[14] Solum (1992), pp. 1238–1239.

[15] Dutch Civil Code (Burgerlijk Wetboek, BW), Book 2, Article 1 and 2.

the understanding of an entity as a rational agent, and the existence of consciousness when it concerns natural persons.[16]

The overlap of the assumptions and the relative priority accorded to each assumption is continuously evolving to accommodate new issues arising time, place and culture. For instance, human slaves in the Roman Empire, as well as in later centuries, were not considered human beings for a long time, nor did they have human rights. They had the possibility of peculium though, to have and hold a certain amount of property as their own private property that their masters allowed them to spend or use as their own. Still, they were considered to be property; legal objects that could be bought or sold. So, we see that in that time legal objects and legal subjects could coincide.

In the U.S., on the other hand, slaves could be punished for criminal acts so as to exclude the criminal liability of their masters.[17] This is comparable with the treatment of animals in criminal law as was common in Europe in the Middle Ages, as will be described below. Among human, so-called natural persons, there has always been a difference in the contents of the legal capacity of legal personality. Also, in "modern times" there existed, and still exists, legal discrimination amongst natural persons. Until recently, for example, women in all western societies were not considered to have comparable legal capacities as their male counterparts. Until 1957, married women in the Netherlands, for example, could not perform legal acts without the consent of their husbands.

Changes continue to take place regarding the legal status of minors and their capacity to perform activities with legal capacity. Rights based on age or gender to drive cars, vote, buy weapons or marry vary per culture, time and place.

In addition, society has allowed the creation of artificial business entities such as the corporation, firm or foundation, based on the necessity that these entities have to have the power and legal status to perform economic acts with legal consequences and have to have legal credibility. In our present society we have discussions and even legal actions to consider personhood for animals. There have also been recent actions granting personhood to inanimate objects such as the Whanganui River in New Zealand and several rivers in India, suggesting that the scope of the legal construct of personhood may be expanding if the need arises.[18]

Whether an entity should be considered a legal person depends on the following question: should this entity be made subject of a specific set of legal rights and duties? The answer depends upon the cultural, economic and political circumstances. There is considerable confusion about this central legal question, as well as deep intellectual divisions.[19] Legal personhood can be considered for humans, animals or inanimate

[16]Ohlin (2005), p. 210.

[17]American law was inconsistent in its constitution of the personality of slaves. While they were denied many of the rights of "persons" or "citizens," they were still held responsible for their crimes which meant that they were persons to the extent that they were criminally accountable. The variable status of American slaves is discussed in Fagundes (2001) and Naffine (2003), p. 346.

[18]Hutchinson (2014).

[19]Naffine (2003), p. 346.

objects if you think of law from an essentialist perspective, as an artificial pragmatic construct, meant to service society. Of course, this also applies to legal objects and all norms translated in laws by humans. Or one could choose the concept of comparatism in the sense of Cartesian dualism. This would entail separating the concepts of legal personhood and legal objects on the basis of their characteristics as consciousness, matter, will etc. To compare these concepts, one could take the common characteristics to find the most applicable legal status for different manifestations of robots or AI driven systems. But, in addition, a complete dualistic principle of the concept of legal personhood is possible based on the functional requirement of legal capacity of the entity concerned as was the case with artificial personhood.

Legal personhood is a flexible and changeable aspect of the legal system. As stated by international lawyer, Ian Brownlie, it is well-recognized that the subjects of law in any legal system are not exactly identical in their nature and rights or in the extents of their rights and nature, depending on the needs of the community.[20] And certainly in international law, the recognition of the responsibility as a legal subject varies and often is used to protect the "legal subject" state to push the other legal subject in front of them:

> There is no international criminal law which applies to states as accused, but there is an increasing body of rules, administered in part by international tribunals, which subjects the conduct of individuals (potentially including state officials) to international criminal law. These developments, particularly in the field of human rights, have added another category of personality (albeit heavily qualified) to those within the international legal system, namely individuals and sometimes corporations created by national law.[21]

Specifically, in international law it is recognized that the scope of legal personality is measured by the need of society under different circumstances.[22]

To see if and how AI-driven entities as autonomous robots need legal personhood, it is helpful to compare them with the bearers of legal rights and obligations that currently exist in our society, namely natural and artificial legal persons. For a more essentialist vision, it is necessary to look at the bare necessity that is essential to the function of the autonomous robot in a more metaphysical way. This is the artificial legal layer—a legal fiction—that can be applied or taken away in the sense of a construct of the "Cheshire Cat," the non-existing entity that can be there if one needs it and vanishes when superfluous.[23]

---

[20]Brownlie (1990), p. 58.

[21]Crawford (2012), p. 17.

[22]"All that can be said is that an entity of a type recognized by customary law as capable of possessing rights and duties and of bringing and being subjected to international claims is a legal person. If the latter condition is not satisfied, the entity concerned may have legal personality of a very restricted kind, dependent on the agreement or acquiescence of recognized legal persons and opposable on the international plane only to those agreeing or acquiescent." Crawford (2012), p. 117.

[23]Naffine (2003).

5 The Physical Person as a (Natural) Legal Person

Starting from Cartesian dualism and combining the sentient and conscious characteristics of human beings to the legal conception of natural persons, it is easy to see that there is a clear separation between natural persons and legal persons. To identify which aspects of legal personality might apply to AI entities as autonomous functioning robots, it is helpful to explain the relevant characteristics of natural and unnatural legal persons. Legally, the individual as a natural person is the bearer of rights and obligations due to the fact that it concerns a living person and not a fictional entity.

However, there is some agreement on what is characteristic of the individual: each individual differs from the other in the physical sense, but in a legal sense, each man of flesh and blood is the bearer of rights and obligations. The law regulates who is Dutch, German, American or Chinese and that everyone in the Netherlands is the bearer of rights and obligations; however, the law does not regulate just what a natural person is.[24]

From a historical perspective we can look at the concept of person and personhood as defined by Thomas Hobbes in his famous work Leviathan. According to Hobbes, a person is:

> He whose words or actions are considered, either as his own, or as representing the words or actions of another man, or of any other thing to whom they are attributed, whether truly or by fiction.
>
> When they are considered as his own, then is he called a natural person: and when they are considered as representing the words and actions of another, then is he a feigned or artificial person.[25]

Hobbes explains the origin of the word coming from the Latin "persona" and the Greek "prosperon," a mask used in theatres.[26] Still the Romans reserve this "persona" phenomenon to living (natural) humans, including women and slaves. The last two though were not to be considered as equal to the natural male citizens.

Hobbes separates the phenomenon of legal personality for non-human actors from artificial legal persons; if the person does not speak for himself, but their action or representation is attributed, one can speak of artificial personality. Hobbes' concept does not necessarily imply that this must be a human. Of course, he did not account for autonomous robots but he might well have considered this if he had been confronted with autonomous and maybe sentient robots. As referred to by Pagallo, the idea that a legal subject can be an "artificial person" should be traced back to the notion of "persona ficta et rapraesentata" developed by the experts of Canon Law since the

---

[24] Article 2 of the Dutch Constitution (Grondwet, GW), in conjunction with the Dutch Civil Code, Book 1, Article 1.

[25] Hobbes (1651).

[26] The word "person" is Latin, instead whereof the Greeks have "prosopon," which signifies the face, as "persona" in Latin signifies the disguise, or outward appearance of a man, counterfeited on the stage; and sometimes more particularly that part of it which disguiseth the face, as a mask or vizard: and from the stage hath been translated to any representer of speech and action, as well in tribunals as theatres. Hobbes (1651).

thirteenth century. And Thomas Hobbes' Leviathan has a precedent in the work of Bartolus de Saxoferrato (1313–1357).[27]

Another feature of the natural person is found in the spiritual aspect of the natural person. In religious scriptures one often finds references to the presence of the soul. Artificial legal persons and objects do not have a soul (whatever that may be). According to the catechism of the Catholic Church, which still can be seen as an expert in this field, the word "soul" is defined as follows: "soul" means the spiritual principle in man. The soul is the subject of human consciousness and freedom.

The freedom of decision is the ethical and legal background of the responsibility we have as natural beings. Individuals are sovereign in their decisions and therefore legally responsible for their actions.

Jean Bodin claimed that sovereignty must reside in a single individual. This sovereignty can be transferred to other legal entities, i.e., the state, a company or any other organizational unit. These legal entities must be considered "legal entities" with the power to make decisions with legal effects.[28]

The important question is whether independent, technical and electronic instruments, combinations of hardware and software or algorithms, can be considered as bearers of rights; whether these might be vested with the power to act as legal entities and thus can perform legal acts, or whenever they are mandated to produce such acts. Their actions could also lead to liability that is not directly traceable to any other responsible body as is the case with employees, children and animals. Or will there always be a natural individual behind the acting entity as the ultimate bearer of the rights and legal responsibilities?

An individual will always be a legal entity with legal personality but a legal entity, an artificial entity will not have the same rights as a natural person. The legal entity, being a natural person, the subject of rights and duties, can act with legal implications. There is no question whether one of these natural persons is fictitious or natural. One is of flesh and blood, but inherent to this person is that he is able to function socially and, if legally competent, able to perform acts with legal consequences.

Natural persons can vote for other individuals in elections and be elected to represent other individuals. They may join a political party or a church. They will be the subject of human rights, the right to life, privacy, freedom of expression, right to education and freedom of religion. Individuals may be put in prison if convicted for a felony. Individuals can marry another person or enter into a civil partnership. They may have children by natural birth from other individuals and will have an automatic natural and legal relationship. And yet, this also changes over time. Due

[27]In his Commentary on Digestum Novum (48, 19; ed. 1996), Bartolus reckons that an artificial person is not really a person and, still, this fiction stands in the name of the truth, so that we, the jurists, establish it: "*universitas proprie non est persona; tamen hoc est fictum pro vero, sicut ponimus nos iuristae*." This idea triumphs with legal positivism and formalism in the mid-nineteenth century. In the *System of Modern Roman Law* (1840–1849) ed. (1979), Friedrich August von Savigny claims that only human fellows properly have rights and duties of their own, even though it is in the power of the law to grant such rights of personhood to anything, e.g., business corporations, governments, ships in maritime law, and so forth." Pagallo (2013), p. 156.

[28]Bodin (1955).

to bio-technology, individuals can also be fully or partially naturally inseminated, derived from insemination with sperm or ova from a third party. It is even possible that children are the result of a DNA merging from three different individuals.[29] For the time being, at least, this has no legal consequences.

Would that situation be different if the DNA is continued to be manipulated? Or, if use is made of non-natural or non-human DNA? Is there a boundary between natural and non-natural persons? What position will the "semi-natural" person have in which robotics and individual intelligence will complement each other?

The law and legal opinions may not give an answer or have a final say on these questions. This is the terrain that legal science shares with other sciences: political science, medicine, ethics, psychology and metaphysics.[30]

Meanwhile, the question remains what features are relevant to determine what a real or natural person is? Bio-technology and AI are converging. Artificial limbs and organs are already integrated in the human body. Also, several experimental couplings of the brain to the Internet of Things has already occurred.[31] Bio-engineering is developing at an incredible pace.

So, the question arises if a natural person can be considered a complete natural person if a considerable part of their body and, specifically, the functioning of this human system is supported by artificial elements. Can a natural person have 50% artificial organs and limbs? Or 75%? Is this relevant from a legal perspective? Do we discriminate on the basis of free will and intelligence? If an individual is not able to independently perform legal acts under curatorship and if the individual is not legally defined as an adult (in the Netherlands and other countries, 18 years), natural persons are not able to perform acts with legal consequences. Mentally incapacitated adults also could be put under curatorship. This is not an absolute rule. Minors and adults under guardianship and can buy a sandwich, ice cream or even a bicycle, but will not be able to buy a car or a house. Their parents or trustees have a duty to support them and to represent them. There is a transition period between full responsibility for the actions of children, to adulthood; which usually begins between 14 and 16 years old, and in China even from 10 years onward. The parent or guardian is not liable if he is not at fault for a harmful act by the child. But even within this system there are cultural and national differences. The age of full legal capacity is well established in The Netherlands and the U.S. at 18 years of age. However, an "adult" person in the USA is not allowed to buy alcoholic beverages, but is allowed to drive a car at the age of 16 or can purchase a fire arm, as referred to above. In many countries in Africa and Asia, for instance, there is no minimum age set for marriage. India recently had the maturity and judgment limit lowered to 16 years of age for the perpetrators of a

---

[29] Hamzelou (2016) Exclusive: World's first baby born with new "3 parent" technique. Available at: https://www.newscientist.com/article/2107219-exclusive-worlds-first-baby-born-with-new-3-parent-technique/. Accessed 11 October 2017.

[30] Geldart (1911), p. 94 and Dewey (1926), p. 655.

[31] Brainternet works by converting electroencephalogram (EEG) signals (brain waves) in an open source brain live stream. Minors (2017) Can you read my mind? Available at: https://www.wits.ac.za/news/latest-news/research-news/2017/2017-09/can-you-read-my-mind. Accessed 11 October 2017.

crime. Thus, the law is far from consistent, not even nationally and certainly not in an international context. Legal standards are not equal for natural persons. There is also a tendency to look at the quality of the psychological capacity of natural persons. An example is the (not accepted) proposal to forbid women to have children when the parents are apparently not able to raise their children adequately, for example if they already have children expelled from home to external care.[32]

Furthermore, reference may be made to the historical context in the perspective of the standards relating to the legal capacity of natural persons. Time and culture vary with the legal status of natural persons. The abolition of slavery and, therefore, the abolition of the (partial) status as legal object only took place in 1794 in France, to be renewed by Napoleon in 1802, finally abolished in 1841 in 1838 in the UK after the abolition Act of 1833. The Netherlands and the U.S. finally accepted the abolition of slavery in 1863. Nevertheless, there are people still living and working under "slave like" circumstances.[33]

Women only got their democratic voting rights across the western world at the beginning of the 20th century. Until the abolition of the law on incapacity on 14 June 1956, married women in the Netherlands were legally incapacitated.[34] Belgium maintained this rule until April 1958. However, until 1971, the Dutch Civil Code stipulated that the man was the "head of the family" and that the woman owed him obedience. To increase the complexity about legal positions, we may also refer to the fact that a distinction is made in the rights of individuals as such. Same sex marriages are still not allowed in a majority of countries. In many countries, homosexuality is still illegal, and, in some countries, even subject to the death penalty. In conclusion, it can be established that the actual content of the legal status of individuals is not homogeneous. The legal status of natural persons is not manifest and is dependent on time as well as social-cultural circumstances. This point of view can also be applied to the legal characterization of the robot.

5.1 Natural and Human-like Behavior as Determination for Legal Personhood

Some legal scholars argue that legal personhood should be limited to human beings or, at least, to serve the legal system that is construed by and used for the benefit of human beings. Their fear is that *extending the class of legal persons can come at the expense of the interests of those already within it.*[35] In the film "Bi-centennial

[32]Proposal Ira Winds, Livable Rotterdam alderman.

[33]Aziz and Hussain (2014) Qatar's Showcase of Shame. Available at: https://www.nytimes.com/2014/01/06/opinion/qatars-showcase-of-shame.html?_r=0. Accessed 12 October 2017; The Global Slavery Index (2016) https://www.globalslaveryindex.org/findings/. Accessed 12 October 2017.

[34]On 14 June 1956 the House settled the bill by Minister JC Furnace, so that married women were legally competent as from 1 January 1957.

[35]Bryson et al. (2017), p. 275.

Man" based upon a book by Isaac Asimov, the robot Andrew Martin wants to be recognized as a natural person.[36] Initially, his request was rejected by the President of the Court because a robot can simply not be recognized as a natural person. The consideration bearing this rejection is that a robot lives forever and cannot die. Years later, the robot, when adapted so that he can die, requests a revision of this judgment.

> Andrew Martin: In a sense I have. I am growing old; my body is deteriorating, and like all of you, intending eventually to cease to function. As a robot, I could have lived forever. But I tell you all today, I would rather die a man, than live for all eternity a machine.[37]

An interesting discussion, especially since various scientific sources suggest that old age is a disease that can be cured and every human being in any case would soon be 130 years old and perhaps even immortal in time.

Another consideration that is used to obtain the qualification of a natural person is the existence of free will: "It has been said in this courtroom that only a human being can be free. It seems to me that only someone who wishes for freedom can be free. I wish for freedom."[38]

The idea of qualifying an autonomous thinking and self-decisive robot as an individual based on the autonomy and free will is a fairly extensive one. Free will, as indicated by Descartes, is based on the fact that we, as human beings, have the experience by which free will steers our behavior. Aristotle had the conviction that this free will also exists within animals.[39] And is not our "free will" determined by circumstances, history and genes? And are we conscious of this free will? Is that consciousness? According to Shaun Nichols in an article in the Scientific American it is just a series bio-electric signals, not more, referring to neurons firing in certain brain areas, no more and no less.[40]

For autonomous thinking there is also the need for intelligence. This aspect is also often used to determine the humanlike behavior, needed to determine the determination of a human and therefore a natural person.

The problem is that the concept of intelligence is not very extensively defined due to the different concepts of intelligence, i.e., rational intelligence and social intelligence. Howard Gardner theorized that there are multiple intelligences comprised of nine components: naturalist, existential, musical, logical-mathematical, bodily-kinesthetic, linguistic, spatial, interpersonal, and intrapersonal intelligence.[41]

---

[36] Isaac Asimov's The Bicentennial Man and Other Stories (1976) and later edited by Asimov as The Positronic Man (1993), co-written with Robert Silverberg, ultimately raises formed the basis for the script of the movie Bicentennial Man, 1999 starring Robin Williams.

[37] Isaac Asimov (1976).

[38] See Footnote 37.

[39] Descartes (1973).

[40] Shaun Nichols, Is free will an illusion? Available at: https://www.scientificamerican.com/article/is-free-will-an-illusion. Accessed 10 March 2018.

[41] Gardner (1993).

David Wechsler formulated in 1955 a well-known general definition of intelligence: "The aggregate or global capacity of the individual to act purposefully, to think rationally, and to deal effectively with his environment."[42]

Without going into the attitudes that exist about the many forms of intelligence, I would limit this reference to the intelligence needed to participate as an individual in society. To this end, it is necessary that there is understanding of the consequences of acts performed in this social traffic (with legal effect). Certainly, AI entities will be capable, now or in the near future, to meet the Turing test, the qualification for intelligence on a "human" level.[43]

Compliance with this test, something that other animal primates certainly cannot meet, gives the impression one has to do with a human being. Yet there are regular attempts to give these other primates a form of legal personality. The chimpanzee, an entity that is regarded as reasonably intelligent, was subject in the appeal court in New York on an appeal to personal liberty (Habeas Corpus).[44] The status as a natural person was not accepted. The court stated that chimpanzees, although cognitively complex, are not entitled to the same legal status as human beings: "We conclude that a chimpanzee is not a 'person' entitled to the rights and protections afforded by the writ of habeas corpus."[45]

Only people can have rights, the court states, because only people can be held legally accountable for their actions. "In our view, it is this incapability to bear any legal responsibilities and societal duties that renders it inappropriate to confer upon the chimpanzees legal rights…that have been afforded to human beings."[46] On the other hand, the court also states that: "the classification of a being or entity as a 'person' is made solely for the purpose of facilitating determinations about the attachment of legal rights and duties."

The Non-Human Rights Project, the appellant in this case, did not agree with the ultimate conclusion of the court and stated:

> The Court ignores the fact that the common law is supposed to change in light of new scientific discoveries, changing experiences, and changing ideas of what is right or wrong; it is time for the common law to recognize that these facts are sufficient for the establishment of personhood for the purpose of a writ of…[47]

---

[42]Wechsler (1955).

[43]The Turing Test published by Alan Turing (1950), was designed to providence a satisfactory operational definition of intelligence. Turing defined intelligent behavior as the ability to achieve human-level performance tasks, sufficient to fool an interrogator.

[44]State of New York, Supreme Court, Appellate Division Third Judicial Department. Decided and Entered: December 4, 2014 (518336). Available at: http://decisions.courts.state.ny.us/ad3/Decisions/2014/518336.pdf. Accessed 20 October 2017.

[45]State of New York, Supreme Court, Appellate Division Third Judicial Department. Decided and Entered: December 4, 2014 (518336), p. 6.

[46]State of New York, Supreme Court, Appellate Division Third Judicial Department. Decided and Entered: December 4, 2014 (518336), p. 5: Amadio v Levin, 509 Pa 199, 225, 501 A2d 1085, 1098 [1985, Zappala, J., concurring] [noting that "'personhood' as a legal concept arises not from the humanity of the subject but from the ascription of rights and duties to the subject"].

[47]The Nonhuman Rights Project (NhRP) further stated: chimps and other select species—bonobos, gorillas, orang-utans, dolphins, orcas, and elephants—are not only conscious, but also possess a

Although Descartes was able to claim that animals are mere machines due to their lack of cognitive abilities, the discussion above has indicated that this vision is slightly impaired. Animals are not "things"; therefore, provisions with respect to issues on animals apply and should be in compliance with the laws, regulations and rules of unwritten law, reasonable restrictions, obligations and principles of law and public order and decency.[48] Although animals still have no rights, they will be treated on the basis of their role in society, yet with certain rights based on the obligations of natural persons in society. Abuse or neglect of animals will not be accepted and rules as such are also included in the Criminal Codes; and certain rights for animals, in the Netherlands since 2011, are included in "the law on animals."[49] This animal has no legal personality but there is a societal tendency to have more rights applicable for animals and not just to the legal and beneficial owner of an animal. The owner and others have been given more responsibilities with regard to the animal in the context of acting carefully and friendly. Animals are not considered to be objects.

The question is whether in respect to certain social robots as pets, companion robots and sex robots, the same regime should apply.[50]

Yet there are also voices to provide animals with some form of legal personality. Animals can take various decisions under the influence of different information. Is this proof that they have a comparable free will that also proof a cognitive base for their decisions on the information obtained?

It is stated that a chimpanzee is not much inferior to man when it comes to their reason, intelligence, temporal insight, self-awareness, self-control, theory of mind, social and emotional repertoire, curiosity, communication and its ability to desire and act intentionally. According to the advocates of rights for chimpanzees, these capabilities ideally are what a chimpanzee makes a person, a legal entity and a bearer of fundamental rights. Unfortunately, for the animal rights advocates, this reasoning has so far not resulted in effective legal personality for animals. Personally, this animal-as-a-person-reasoning often tends towards wishful thinking. Not many people would, for example, trust a trained chimpanzee behind the wheel of a car if the car is under that chimpanzee's control. Nor is it particularly uplifting to enter into a conversation with a chimpanzee. The probability that meaningful communication will be possible with an intelligent robot, at least at a higher level, is considerably larger.

---

sense of self, and, to some degree, a theory of mind. They have intricate, fluid social relationships, which are influenced by strategy and the ability to plan ahead, as well as a sense of fairness and an empathetic drive to console and help one another. In many ways (though certainly not all), they are like young children. The NhRP contends, based on this, that chimpanzees are capable of bearing some duties and responsibilities.

[48] Dutch Civil Code, Book 3, Article 2a.

[49] Article 350 paragraph 2 of the Dutch Penal Code (Wetboek van Strafrecht) and Law of May 19, 2011, on an Integrated Framework for Regulations on Captive Animals and Related Topics (Animals Act).

[50] Darling (2016).

5.2 *Non-natural (Artificial) Legal Persons*

The non-natural person may be a company or other entity such as an organization, institution, a foundation, etc. An individual cannot be a business but will represent a legal entity to act with legal force. In more structured societies it was common to use the entity of the corporation. In ancient Egyptian society the legal structure of a foundation was used to maintain temples. In Roman civilization, there were several legal entities such as the "universitates personarum," which was similar to a corporation or government college with their own identity and independent legal personality.

A well-known Dutch international organization with legal personality, the first multinational corporation, was the Dutch East India Company (VOC), founded in 1602. This is another clear example of adapting the legal reality to the social and economic needs of the times.

A legal person, as to property, is in an equal position as an individual and natural person, unless otherwise provided by law. A legal person is, in a similar way to the individual, a legal entity to participate in socially relevant legal relationships. A legal person can go to court if its interests are affected, or can be sued in court if it acted unlawfully in the view of another legal or natural person.

As John Dewey indicated already in 1926 in the Yale Law Journal: "The Corporation is a right-and-duty-bearing entity."[51]

As stated, corporations are not equal to humans, but they do have a legal personality to act in a legal sense. Although it is a legal fiction, granted to organizations and other entities they can only act in a legal manner that is in the best interest and the purpose of this legal entity. Thus, the fiction is a kind of augmented reality as a legal layer to social reality and not imaginary, at least not within a society that is based on a legal reality. There is a global spectrum of legal persons in civil law. In the U.S., this means even to some extent, the application of the Bill of Rights guarantees to corporations. Carl Mayer describes this situation in the United States on the basis of the development of equal treatment under the Fourteenth Amendment. Companies are considered persons for the purpose of the Fourteenth Amendment, i.e., companies should have the right to equal protection and due process.[52] Of course, these conceptions are not equally applied across the globe. As stated before, legal as well as social conceptions differ throughout countries, cultures and political structures.

Can we derive useful comparisons from these characteristics to define a legal framework for the artificial intelligent entity?

---

[51] Dewey (1926), p. 26.

[52] Mayer (1990).

6 The Autonomous Artificial Intelligent Robot

For the sake of this section, considering the legal status, the robot is assumed to be an autonomously functioning artificial intelligent self-learning entity. AI is described as a system applied to an advanced computer technology, which is aimed at imitation of intelligent human behavior,[53] partly to understand (human) intelligence and also to create intelligent creatures that can operate autonomously in complex, changing situations.[54] Will such a system need legal personhood? This will depend on the dimension where it will function, in society, in culture and its intended purpose. For instance, as applied to robots with multipurpose tasks that require intelligence and social behavior, a certain legal competence is thinkable. As the possible cooperation between those autonomous robots and natural persons will be very probable a legal mutual commitment based on trust is a prerequisite.

This line of thinking is also observed in the earlier referred to the motion of the European Parliament in consideration 50:

> Notes that development of robotics technology will require more understanding for the common ground needed around joint human-robot activity, which should be based on two core interdependence relationships as predictability and directability; points out that these two interdependence relationships are crucial for determining what information need to be shared between humans and robots and how a common basis between humans and robots can be achieved in order to enable smooth human-robot joint action…

Of course, this moment is still shrouded in the nebulae of the future but it is probably nearer than we think given the pace of technological developments in this context.

6.1 *The Increasing Use of AI in Robotic Entities*

The example of a semi-autonomous functioning system is IBM's Watson as it carries out numerous tasks at the moment in the field of DNA research, teaching and seed breeding, to name a few.[55] Nevertheless, this system still receives its initial instructions from an individual. Even under these limiting circumstances, one could consider that there are certain legal effects that result from its own functioning. This could provide for certain attributed legal personhood, be it that there have to be limits to the extent of legal consequences, as will be explained later.

In today's society such systems or robots are still (at least partly) controlled by natural persons. However, there is an undeniable trend towards the use of self-thinking and self-acting systems. Also, natural persons are controlled in their professional

[53] Shoyama (2005), p. 129.

[54] Russell and Norvig (2010), pp. 1 and 18; also referring to the following definition of AI: The act of creating machines that perform functions that require intelligence when performed by people. Kurzweil (1990).

[55] See http://www.ibm.com/watson/.

activities in comparable ways by other natural persons or (artificial) legal persons. AI applications will be in the field of all kinds of industries, such as hosting, social and physical support, care robot in physical and social sense, the sex robot, industrial robots, medical robots, surveillance robots, military robots, drones, etc. In the medical sector molecular nano-robots are deployed of chemical or organic origin.[56]

The fear of the unknown creeps up on us when AI becomes uncontrollable in the sense that we cannot understand the processes that move the AI system or entity because the self-learning and teaching element is beyond our human comprehension. This is the so-called "super intelligence" and is the result of the singularity based on Moore's law and paradigm shift. Moore observed the fact that the capacity of microprocessors doubled every two years. Vinge and Kurzweil broadened this concept to other technological developments, including a (paradigm) shift to other forms of technology if the former development would hamper the further progress, for instance from micro-processing to nano-processors.[57] This increase would also manifest itself in the development of intelligence by artificial means, resulting in super intelligent entities of a bio-digital character or, of course, a manifestation not yet known to mankind.

Nick Bostrom has defined super intelligent systems as: "Any intellect that radically outperforms the best human minds in every field, including scientific creativity, general wisdom and social skills."[58]

It is alluring to elaborate further into the apocalyptic scenarios predicted by Vinge and Bostrom and others but I will restrict myself to the legally relevant perspective. The robot is not yet super-intelligent but can be considered as a dynamically evolving concept that started as a machine, fueled with AI and is constantly evolving into a complex autonomous functioning robot and—maybe in a later stage—super-intelligent or semi-humanoid system.[59] The nature of this entity—electronic, or organic-chemical—is less relevant for its legal characterization. The state of intelligent autonomy and its function in society will be more relevant in determining its legal status.

One could refer, in this respect, to the development of the "intelligent" car. This is already happening and therefore an understandable example. The modern automobile is quickly developing an increasing autonomous mode of operation. We already

[56]Examples are the molecular machines as designed by prof. Ben Feringa, Nobel laureate in 2016.

[57]A specific paradigm (a method or approach to solving a problem, e.g., shrinking transistors on an integrated circuit as an approach to making more powerful computers) provides exponential growth until the method exhausts its potential. When this happens, a paradigm shift (i.e., a fundamental change in the approach) occurs, which enables exponential growth to continue. Available at: http://www.kurzweilai.net/the-law-of-accelerating-returns Accessed 10 May 2018.

[58]Bostrom (2014).

[59]Already in the 1960s this development was predicted: let an ultra-intelligent machine be defined as a machine that can far surpass all the intellectual activities of any person, however clever. Since the design of machines is one of these intellectual activities, an ultra-intelligent machine could design even better machines; there would then unquestionably be an "intelligence explosion," and the intelligence of humans would be left far behind. Thus the first ultra-intelligent machine is the last invention that humanity need ever make, provided that the machine is docile enough to tell us how to keep it under control. Good (1965), cited by Vinge (1993).

drive with all kinds of warning systems, automatic breaks, distance keeping mechanisms, etc. According to the road traffic law, the driver is the responsible party. But how to justify this when the driver is gradually losing control over the car and, instead, depends on numerous providers of information? These providers are the manufacturer, the infrastructure, road managers, other motorists, the producer of the software, the meteorological department, the designer of the algorithm at the heart of the learning vehicle and third-party data providers that control or affect navigation and engine control. Therefore. the suggestion in the earlier mentioned European Parliament motion to adapt the outdated Vienna Convention on Road Traffic of 1968 is not undue.

> 10. Expects the Commission to ensure that the Member States adapt the existing legislation, such as the Vienna Convention on Road Traffic of 8 November 1968, in a uniform manner in order to make driverless driving possible, and calls on the Commission, the Member States and the industry to implement the objectives of the Amsterdam Declaration as soon as possible.[60]

But what if a direct link between the brain activity of the "driver" and the software control is made? Not so futuristic, there are already cars that respond to drivers who threaten to fall asleep where certain movements betray a delay in reflexes. Going one step further, those links are analyzed in an external autonomous system that will control the traffic flow. This not the plot of a science fiction novel. Elon Musk is also moving into neuro-tech; he launched Neuralink, a company that is researching methods to upload and download thoughts. Ultimately, Neuralink aims to change the way in which we interact with devices by linking our brains to the machines we interact with most often: cars, mobile devices and even smart items in the smart home of the future. This also is happening in the academic research at the University of Witwatersrand, SA as referred to earlier: the "Brainternet" project streams brainwaves onto the Internet. Essentially, it turns the brain into an Internet of Things (IoT) node on the World Wide Web. IoT refers to connecting any device with an on and off switch to the Internet."[61]

Another example is an AI application that is used in the selection of candidates for jobs. Beyond the algorithmic selection of candidates based on their email or letter, by so called Applicant Tracking Systems (ATS), AI can evolve into AI robots that can be used during a conversation to watch an individual's posture, eye movements, sweating, tuning stability and other mental and physical reactions. This analytical achievement will be developed to even a greater extent in the "care-industry," where autonomously functioning robots will apply client custom made solutions to the needy without the necessity of guidance from outside.

To determine the legal classification of the AI entity as a simple tool, a legal object that is used as an instrument; or as an autonomous artificial intelligent entity that will operate independently and could be classified for legal activities, we have to

---

[60]Reference to the Declaration of Amsterdam of the Council, of 14–15 April 2016, on cooperation in the field of connected and automated driving ("Amsterdam Declaration").

[61]Minors (2017) Can you read my mind? Available at: https://www.wits.ac.za/news/latest-news/research-news/2017/2017-09/can-you-read-my-mind. Accessed 11 October 2017.

determine its role and status.[62] Whether robots should be compared to legal persons or legal objects is to be answered for a great deal on the basis of function and autonomy. This decides whether they are assessed similarly as thing as minor, non-subordinate, as movable property and animals,[63] or as independent legal entities.

A complicating factor is that it is not so easy to tie to a breakdown of legal persons and legal objects. The artificial legal person that is a company can be an object too, it can be sold and it can be divided; but also, it can be held responsible for its actions. To determine a sensible solution for a new legal personhood structure, if needed, we have to develop an original analysis.

7 The Question of Punishment of (Legal) Persons: A Criminal Law for Robots?

Looking with simplifying glasses at the criminal law one can regard it is as an instrument given to the state by its subjects or obtained in a less democratic way by a state authority, with the purpose to secure law, order and security in the society. The content is directed at the offender of these specific societal rules of behavior and social values and norms and consists of punishment of this behavior with the intention to punish, correct or re-socialize the offender. This system is developed to keep human behavior between the lines of society but, of course, is dependent on the time, in the sense of the era, culture and political system.

An exemplary issue for a lot of people contemplating the legal difference between the legal position of robots and humans is the question how to punish a robot if "it" commits a crime. Also, scholarly colleagues often ask me in what way we should punish robots if they would commit a crime, as this is a pitfall to give up the legal positioning of robots. Of course, the question is easier to state than the answer.

It will be dependent on whether we accept the robot as a legally and morally accountable entity, a legal subject, or just as an object. The question is whether this legal description will suffice for a clear separation. Loyal to the comparison with other mammals and, in particular, with human beings as well as artificial legal persons, we have to start the comparison with these "structures."

Comparison with existing legal persons only suffices if we want to connect to the ideas of the positive legal system of criminal law where there is a strong conviction that the deed has been committed by natural persons or, at least, under the

---

[62]The Principles of European Tort Law ("PETL") refers to liability for "auxiliaries" (6: 102)—an apt term for both robots, although in PETL it is meant particularly for people. Article 3: 201 of the Draft Common Frame of Reference (DCFR) of the Principles, Definitions and Model Rules of European Private Law refers to workers or "similarly engaged" others, in which the phrase "similarly engages" others may contain cases of accidental damage; see: Giliker (2011), pp. 38 et seq. Then the robot will have to be seen as "another," where the employer is liable under the condition that he still has "the least abstract possibility of directing and supervising its conduct through binding instructions"; Von Bar and Clive (2009), pp. 34–55.

[63]Schaerer et al. (2009), pp. 72–77.

responsibility of natural persons as in the case of artificial legal persons. After all, companies can commit crimes. These crimes are mostly of a financial character, such as fraud, money-laundering or tax crimes. But, also environmental crimes involving pollution by chemical and oil industries, or false reporting as in "Diesel gate" in the automobile industry and even discrimination of clients or in the personnel area. Mostly, the punishments are fines, sometimes extremely high if it considered to be a crime against competition rules, for example. Also, states can commit crimes as polluters, financial villains or war-criminals. Generally, such crimes will be paid out of the financial reserves of the company and, in rare cases, the responsible board members, or in the case of war, responsible state commanders, will be put on trial.

It is not unimaginable to submit entities other than human beings to the criminal law. As alluded to above, in the Middle Ages, several criminal proceedings were held against animals in the same way as they were held against humans. In 1266, in Fontenay aux Roses, for example, a pig was convicted and brought to death in the city square because the beast had bitten and killed a child. The judge ordered the executioner first to cut off a paw to be followed by a beheading. Before the execution, the pig was dressed in the clothes of a human being. There were other cases against horses, cows and bulls that wounded or killed humans or other beasts. In such cases, the animal was punished without holding their masters accountable. The animals also had the rights of legal support by an assigned counselor.[64] The last case against an animal in the Netherlands was against a bull in the town of Zwolle in 1664 after he impaled someone. His counselor could not do much to save his client; he was stoned and buried alive.[65] Most interesting for the parallel with the robots is that the owner was not held legally responsible.

8 A Different Construction of Personhood

As made clear above, the concept of legal personality or personhood is a very flexible concept. It is all dependent on what is deemed acceptable within a certain society. If animals are accepted to have a certain status in that society and culture, they can have a legal status going beyond that of a mere object. If a company has legal personhood because it is socially and economically desirable, why should it not be acceptable and even desirable to give a robot a certain legal status and to have a new kind of personhood. This practical view of legal personhood is made by Naffine. To see if this analysis will be of help to determine what legal position could be applicable to AI entities one may consider this model. Naffine gives three possible models for legal personhood:

The (lucid) Cheshire Cat;
Any reasonable human creature;
The responsible subject.

[64]Erven D onder de Linden en zoon (1831), pp. 201–203.

[65]"If a bull gores a man or woman to death, the bull is to be stoned to death, and its meat must not be eaten. But the owner of the bull will not be held responsible."

8.1 *Abstraction of the Robot by the Cheshire Cat, Reasonable Human Creature and Resposible Subject Model*

The first definition of the legal person that Naffine discusses is named the Cheshire Cat.[66] According to this definition, to have personhood means nothing more than the formal capacity to be a carrier of legal rights and duties.[67] There is no moral or ethical dimension to this definition. The person exists only as an abstract capacity to function in law, a capacity which is endowed by law because it is convenient for law to have such a creation.[68]

Anyone or anything can be considered a person in the eyes of the law, because the only reason that legal personhood exists in the first place is because of the practical advantages of such an attribution. This definition of legal personhood is the most comprehensive definition of personhood of the three.

This model does not have any moral, ethical, historical or empirical content.[69] Following this definition, there is no reason why animals or other legally functioning entities should not be considered persons. As long as they are able to carry solely one right or legal duty there is no reason to not grant them personhood, even if a human is necessary to enforce that right.[70] This should not be a problem since the same enforcement from a legal competence is required for minors and other legal incapacitated persons. The interesting part is that there should be no requirement for the scope and contents of the legal subjectivity.

This theory also denies the necessity of differentiating between natural persons and artificial persons, or other entities. In either case, the concept of personhood is an abstract concept; neither the natural person, nor the artificial person is more real than the other. Both of their legal personalities are based on the fact that they retain a particular bundle of rights and duties. This is the essence of the Alice in Wonderland character of the vanishing figure: take away the rights and the duties of the person and its legal personality vanishes like the Cheshire Cat.[71] Supporters of this theory thus envisage the concept of legal personality as an empty slot that fits anyone or anything.[72]

Of course, this concept leaves open the question if other legal and natural persons are willing to perform legal actions with this "new Cheshire Cat." This point can be illustrated by the development of robots when this development reaches a point where robots and people look very much alike and almost cannot be distinguished. The concept "uncanny valley," introduced by Masahiro Mori, is used to indicate the

---

[66] Naffine (2003), p. 350.
[67] See Footnote 66.
[68] Naffine (2003), p. 351.
[69] See Footnote 68.
[70] See Footnote 68.
[71] Naffine (2003), p. 353.
[72] Naffine (2003), p. 356.

point when feelings of aversion eeriness to humanoid robots arise.[73] This is when human robots appear almost, but not exactly, like real human beings. The question arises if humans want to create sentient robots that resemble human beings so much, also considering the legal status of robots, giving them rights that reflect their human-like status.

The second concept Naffine proposes is that a (legal) person is any reasonable human creature.[74] Simply put: to qualify as a legal person, one has to be human. This perspective is the most dominant and comes closest to the common language usage of the word person, at least from an Anglo-American perspective. It is common legal knowledge that someone, in this context meaning a human person, becomes a legal person at the very moment of being born or conceived, depending on the legal order and it certainly ends at death. Furthermore, there is the possibility to limit the scope of personhood if the rationality or psychological stability is not present but personhood, as such, still exists.

There are two ways in which this common legal knowledge is interpreted. Firstly, this reasoning could refer to a human being who has been born alive and has not yet died, and is thus considered a human, therefore a person. Secondly, it could refer to the rights and duties of a person that starts to exist as soon as someone is born as a human being and which cease to exist as soon as this same person dies.[75]

Either way, personhood is linked with both biological and metaphysical notions of humanity. Taking this definition, personhood is not a purely legal matter anymore, but concerns instead the question of what it means to be human.[76] This is also the main criticism of this theory from the perspective of the Cheshire Cat definition. Supporters of the concept of the person as a rational human are, according to supporters of the Cheshire Cat concept, misguided because of their reliance on extra-legal biological or moral considerations.[77] The terms "human being" and "person" are being used synonymously and interchangeably by supporters of this second theory.[78]

The definition of the legal person as a human being has the advantage of simplicity. For someone to be considered a person, one does not require any quality except for that of being a human. Therefore, this theory includes all humans, regardless of their mental or physical state, thus being compatible with the human rights movement. In the meantime, this definition excludes—in line with the common legal view—other non-human animals from personhood. Corporations as artificial legal persons, are able to carry personhood under this definition because they are reducible to the relations between the persons who manage them, own them, work for them and

[73] Mori (2012) The Uncanny Valley: The Original Essay by Masahiro Mori. Available at: https://spectrum.ieee.org/automaton/robotics/humanoids/the-uncanny-valley. Accessed 15 October 2017.

[74] Naffine (2003), p. 357.

[75] See Footnote 23.

[76] See Footnote 23.

[77] See Footnote 23.

[78] Naffine (2003), p. 358.

act in mandate.[79] This definition of personhood, however, is not compatible with the demands of the qualification of differences based on the legal requirements by society. It should, however, be considered in giving legal status to AI entities in the same way the artificial legal person is considered as a vehicle for inter-human legal relations and therefore is served with legal capacity.

The third concept of legal personality observed by Naffine, is the rational, responsible actor; a high-threshold definition since not all humans possess the qualities to be considered persons under this definition.[80] This definition insists on a certain level of mental capacity and therefore excludes young children, mentally incompetent humans and animals.[81] This theory recognizes the human form of personhood, but does not see this as the critical characteristic that sets a human apart as a person. Rather, the rationality, the mental attributes and the ability to comprehend a certain situation that determine this situation.[82] Although, seeming to set this definition of the person as the ideal legal actor, it also encounters the danger of elitism. Moreover, the idea is not very original. Most legal orders already have a system of legal incapability in a private and criminal law sense. Naffine states that under this definition, the person can actually be meaningfully subjected to legal punishment for criminal acts.[83] Criminal law has to treat the person as a responsible actor with a free will because otherwise one cannot take responsibility for one's actions. If a person is not capable of making rational decisions, then what is the point of punishing this person? This reasoning already is applied in many legal systems as "being not accountable for one's actions due to psychological stress or other mental or physical factors." One of the main goals of punishment in criminal law is the prevention of a person committing the same criminal offence again. If a person is not capable of making rational decisions in the first place, then they cannot be expected to learn from their punishment. Nonetheless, in the case of criminal law, this definition of a legal person is simplifying reality; in many ways the law shows awareness of the weaknesses and dependence of human individuals and in many ways the law does not require persons to be as rational and responsible as this Naffine definition requires a human to be.[84]

8.2 AI Entities and Robots in the Theory of Naffine

According to the Cheshire Cat theory, humans can allocate legal personhood to anything, regardless of the nature of the entity that it is allocated to.[85] Inanimate

[79]Solum (1992), p. 1239.
[80]Naffine (2003), p. 362.
[81]Naffine (2003), p. 364.
[82]See Footnote 81.
[83]See Footnote 81.
[84]Naffine (2003), p. 365.
[85]See Footnote 68.

entities have been the subject of legal rights at various times in the past. As mentioned above, temples in Rome and church buildings in the Middle Ages have been regarded as persons in the past.[86] So have ships, an Indian family doll[87] and Indian and New Zealand rivers.[88] And certainly a parallel can be drawn with business corporations and with government entities.[89]

As we zoom in on the example of corporate personhood, we can see a lot of parallels with electronic and AI entity personhood. Similar to a corporation, the aims of an AI entity robot may lie in economic profit for the producer or owner of a robot, or in the social welfare of a society. For example, a robot working for an automobile manufacturer may improve production and thus profit for the manufacturer, whilst a robot caring for an elderly person will be carrying out a civic service. The reason why personhood has been invoked for corporations and robots seems to correspond as well; they reduce the responsibility and liability of the owners in case of damage inflicted by the corporation or the robot. Corporate personhood has seen the liability of its shareholders limited to a certain extent by corporate legislation. Electronic persons could fall under similar legislation, with self-teaching, initiative taking AI.

Taking this definition as our base, there should be no problem in granting personhood to AI considering their specific task or function.

Concerning Naffine's concept of legal personality being connected to the human, granting personhood to AI would be a problem. If personhood can only be granted to humans purely based on the fact that they are humans then it would not be possible for AI to obtain legal personhood. Then how is it possible that corporations are granted personhood? But the legal connection to the natural person could be the trait d'union. The property of a corporation is eventually the property of its shareholders.[90] Damage done to a corporation would directly injure natural persons.[91] As such, corporations are reducible to the relations between the persons who manage them, own them, work for them and so forth.[92] So, the fact that corporations have legal personalities does not necessarily mean that AI entities should be granted legal personality or the same legal capacity. The question lingers though if existing legal persons could represent legal persons (and/or natural persons) in the same way natural persons function in representation or in the use of mandates. Could the attribution of rights be compared with those attributed to natural persons although they would not have the same status as natural persons?

Rejection of the human being personhood concept, granting personhood to AI, is based on the conception that acceptance would undermine the meaning of being a person because it reduces the exclusive belonging of personhood to humans. This exclusivity has been represented by religious texts such as the Bible: man is separate

[86] See Footnote 79.

[87] See Footnote 79.

[88] Safi (2017).

[89] See Footnote 79.

[90] See Footnote 79.

[91] See Footnote 79.

[92] See Footnote 79.

from nature and is created in God’s own image. This hierarchy sets humans above “things,” be it animals, property, or the environment.[93] This argument against granting personhood to AI seems to only be problematic if one uses the terms human being and person synonymously and interchangeably. Electronic or robot personhood does not have the intention to interfere with the exclusivity of humans’ place in the world. According to the common legal view, a natural person (being a human) is different from a juridical person. A legal person does not have to be made up out of blood, flesh and DNA, but exists to ease economic traffic and proceedings in a court of law.

Another argument against granting personhood to robots which aligns with this second definition of personhood is that, because of the special place that humankind has granted itself, it is not in the interests of humankind to grant robots personhood.[94] This argument shows similarities with slave-owners stating that slaves should not have constitutional rights simply based on the fact that it is not in the interest of slave owners to grant them such rights and also deny them a comparable human status.[95]

Overall, robots do fit in with this second definition of the legal person with at least some difficulty and bending of the concept. Even though most arguments against the granting of personhood to AI entities can be put into a practical perspective, in which such legal personality may be pragmatic and desirable, robots lack the ultimate aspect which needs to count as a person in the view of the supporters of this theory: humanity in its widest and non-legal sense.

Returning to the person as the responsible, rational actor.[96] The human form is not the critical characteristic that makes a legal person; the rational, mental attributes and ability to comprehend a situation will suffice to be defined as a person. These characteristics will make a person able to have full legal responsibility and to handle in a single capacity in its own right. In the current technological situation, robots are not (yet) able to perform as a legal person under this definition; it cannot act as the fully responsible and capable person that this theory prescribes it to be; robots are still too dependent on humans as they are not fully autonomous and sentient yet. But this can change rapidly.

However, we do not know how the future will unfold. Imagine a possible future in which humanoid AI walks around the globe with great mental capacity, able to comprehend its own situation and have responsibilities[97]; would this sort of robot qualify as a legal person within this definition?

The definition of the responsible, rational actor presumes the presence of a consciousness. Is this prerequisite for personhood something that robots could actually

---

[93] Lovejoy (1936).

[94] Solum (1992), p. 1260.

[95] Solum (1992), p. 1261.

[96] See Footnote 80.

[97] See, e.g., the robot Sophia, of Hanson robotics, <iframe width=“854” height=“480” src=“ https://www.youtube.com/embed/wEqhGVxd6TE” frameborder= “0” allowfullscreen></iframe> and consulted 1–10–2017 (compare “Ava”: Bush, E. (Producer), & Garland, E. (Director). (2014). *Ex machina* [Motion Picture]. United States.

obtain?[98] We do not have a clear notion of what consciousness actually is and so there is little to say about questions that go beyond our basic intuitions.[99] It could be that we cannot only get consciousness out of neurons but also out of transistors. It might as well be that we cannot get consciousness out of anything except neurons, and that we will never be fully able to reproduce it.[100] If robots would be able obtain a consciousness, and then according to this definition, there should be no problem granting personhood to robots. How would the consciousness of this AI be established? Since we do not have direct access to another person's mind, one can only assume consciousness based on behavior and self-reporting.[101] It might be that the artificial intelligent entity claiming personhood would do this on the basis of having a consciousness but would merely be faking its consciousness.[102]

An objection against granting legal personhood could be that robots lack any sort of feelings.[103] But even that could be developed in future AI, by humans or by AI itself. In the context of the legal person as the responsible, rational actor, this characteristic could actually be beneficial for the granting of personhood to AI. Supporters of this theory state that man should be a rational animal, and requires that he should exercise a reasonable control over their passions.[104] As stated before, the criminal law system takes this actor as the ideal legal person.[105] A form of intelligence completely lacking feelings does not have to control its feelings because it does not have them in the first place.

Considering that a robot is not at a level yet in which it could function as a responsible, rational actor, robots cannot be granted personhood under this definition. Granting personhood under this concept in the future depends completely on how successfully AI will develop sentiency in robots. If AI performs in robots as a humanlike consciousness and could therefore act as the responsible rational actor this definition requires it to be, then this personhood could encompass AI.

8.3 Conclusion Concerning the "Naffine" Analysis

To answer the question: "To what extent does artificial intelligence fit in with the different notions of legal personhood?" concerning the three main definitions of legal personhood by Naffine, the following conclusion can be drawn.

i. *The Cheshire Cat.* Supporters of this concept state that legal person status is nothing more than the abstract capacity to function in law and does not have any

[98] Solum (1992), p. 1269.
[99] Solum (1992), p. 1264.
[100] Solum (1992), p. 1265.
[101] Solum (1992), p. 1266.
[102] See Footnote 101.
[103] See Footnote 98.
[104] See Footnote 81.
[105] See Footnote 81.

further requirements except from the fact that people want to grant personhood to an entity. Autonomous robots seem to be easily compatible with this definition of the legal person as it does not require any further qualities whatsoever. The comparison with corporate personhood, as formulated in the EU motion, is a striking one. Both the aims and reasons why personhood for both entities should be invoked, correspond.

ii. *The legal person as a rational human*. Given the personhood of corporations, also a non-human entity, one could argue that AI could have legal personhood too. Supporters of this theory, however, do not agree. Corporations have legal personhood because of the fact that the relationships that govern a corporation are eventually reducible to humans. Therefore, corporations do not lack the key component of humanity which supporters of this definition require for being a legal person, whilst artificial intelligence does. Supporters of this theory could argue that, by granting person status to autonomous robots, we would be undermining the human component that legal personality has. This objection is only a problem if one assumes that the terms "person" and "human" are synonymous and interchangeable. The common legal view is that there is a distinction between natural persons and juridical or legal persons and therefore the granting of personhood to robots does not have to lead to damage of the person status of humans. However, purely looking at the definition of the legal person as a human being, robots would not be able to obtain a person status.

iii. *The rational, responsible actor*. According to this theory a legal person is rational, responsible and aware of its own situation. Thus, this definition excludes young children and other mentally and therefore legally incompetent entities from being considered legal persons. Robots, at least for the moment, do not qualify as a person taking the requirements of this definition into account. But future, more developed AI generated robots could fall under this definition. To be qualified as a legal person under this definition would mean that it would require increased mental capacity, responsibility and a consciousness. Assuming that in the future all these requirements could be met, robots could be granted personhood under this definition.

9 The Artificial Intelligent Entity or Robot as Legal Actor

The foregoing discussion of legal personality needs to compare the role and personality aspects of robots and other AI systems with existing legal personhood or at least with elements of existing personhood. Would having legal personality be desirable for robots and society?

The consideration that such an autonomously functioning artificially intelligent robot should have a secure legal subjectivity is dependent on the actual social necessity in a certain legal and social order. In other words, will a future society still function without any form of legal personality for autonomous artificially intelligent entities? Or will it have a need to place the entity within the framework of legal personhood?

The deployment of autonomous robots in the near future could be comparable to the efforts of individuals representing institutions and organizations, and to the efforts of individuals working as mandated legal representatives. As an example, I refer to a social service that uses a care robot deployment in support of the needy. The robot is capable of managing the household, ordering products and services, conducting physical support and analyzing medical problems and then even performing medical procedures.

The legal consequences of this development are great. A society that depends on autonomous systems and robots cannot do without a legal framework integrating this development. It is quite conceivable that there is a need, in this future society, for a degree of legal responsibility and legal personality of robots so that the legal consequences of such acts can have a place in the legal framework. A distinction needs to be made between fully autonomous functioning entities and those entities that operate on the basis of previous entries by legal persons. Although the "Cheshire Cat" structure seems to be too simple, not considering all social requirements that would be necessary to perform acceptable roles and to be recognized by other legal persons, we can specify the role and function and legal effect of the AI entity.

Furthermore, the development of self-learning algorithms should be embedded legally before proceeding to the question whether legal personality provision to robots is at order.

9.1 *Legal Subject or Legal Object Specialist?*

The definition of a legal subject does yet not completely coincide with the characteristics of an AI entity, but shows an increasing number of interfaces. Because of the variation in types of AI entities, from vacuum cleaner to sex robot, it is impossible to provide a uniform legal regime for robots. But the same goes for legal persons such as limited companies, foundations, etc. These entities are classified by purpose and function and also have different rights and obligations. For individuals there is a similar specification with regard to the act. Children under guardianship as such have a legal status that falls under the supervision of another natural or legal person. But also, individuals will function under supervision or independently and their activity affects their interpretation of legal personality and the performance of their acts. Government officials, secret service officials, the military, but also medical physicians and journalists have a different legal status from other individuals concerning their function and use of rights in society.

As a classification of the specific robots would be desirable, a reconsideration of the degree of legal subjectivity is needed. The legal subjectivity and derived legal capacity need not be equal to the legal personality such as we know it in positive law. The possible extension of legal capacity could be based partly on the concept of existing legal personhood, leading to a new "sui generis" construction, based on elements of legal autonomy for the purpose of the functioning of the robot in society. In this context, a comparison with the "peculium-like" requirements as restricted liability could be of help.

This reasoning applies when it is possible to figure out who the user or owner of the system is, and when there is general acceptance about the responsibility for the system. In the future this will become an increasing problem as systems function more autonomously and interact with similar systems. Car manufacturers of smart cars until now have still accepted a risk liability. This means that the producer accepts responsibility for errors or incomplete functioning of the system and of automatic control systems. But this system may easily come to an end because of the technical and financial burden.[106]

Is the boundary between legal subject and legal object always clear? Legal objects can be goods, services, rights or objects that are the carrier-subjects of rights and obligations. Objects can never be bearers of rights and obligations similar to a legal entity. The legal property concerns, in particular, business, products and services, but is also applicable to more artificial legal person concepts like an organization or company. The lastly mentioned legal persons may perform as a legal object but are themselves legal entities. This special construction is also described as a set of active and passive proprietary elements. The sui generis construction for AI can take this in consideration. Robots could be considered either as objects or subjects depending on the legal activities of other legal actors. One could interact with AI entities with legal effect but the owner also could sell them or pawn them.

9.2 *Liability and Legal Subjectivity*

The liability of a legal person shall also apply to the director or directors, being natural persons at any time during the lifespan of the liability of the legal persons if they had the responsibility or were authorized to act for the legal person. This seems to apply to AI and robots as well. Robots can be classified simply as legal objects, but they can also occupy a special position. In several publications, the comparison has been made with slaves. As also referred to by Ugo Pagallo, Norbert Wiener compared robots with slaves: "the automatic machine, whatever we may think of any feelings it may have or may not have, is the precise equivalent of slave labour." Also referring to Leon Wein in *The Responsibility of Intelligent Artifacts* (1992), in the sense that: "As employees who replaced slaves are themselves replaced by mechanical "slaves," the "employer" of a computerized system may once again be held liable for injury caused by his property in the same way that she would have if the damage had been caused by a human slave."[107] What is more, Voulon stated that the intelligent agent, such as a software robot, was compared with a slave, deployed

[106]Volvo press release (2015) US urged to establish nationwide Federal guidelines for autonomous driving. Available at: https://www.media.volvocars.com/global/en-gb/media/pressreleases/167975/us-urged-to-establish-nationwide-federal-guidelines-for-autonomous-driving. Accessed 20 October 2017.

[107]Pagallo (2013), p. 3 [referring to Wiener (1950)].

to carry out a particular task.[108] We can easily draw parallels with existing machines that perform the needed legal actions to fulfill legal statements and transactions:

> Such a machine would need to have two abilities. First, it must be able to render correct outputs from given factual inputs. Second, its output needs to be reified some way in the real world. The vending machine is the archetypical example of a self-executing smart contract. Vending machines have been defined as 'self-contained automatic machines that dispense goods or provide services when coins are inserted.'[109]

In other words, the vending machine completes one side of a contractual relation. A funny example in this respect is the case of the British bookseller, Richard Carlile, in the year 1822, who invented a book-dispensing machine so as to avoid prosecution under the country's libel and sedition laws. He had been jailed previously and wanted to avoid any future liability, so the idea was to make it impossible for the Crown to prove that any individual bookseller actually sold the blasphemous material. He argued that it was purely a contract between the buyer and the machine with the publisher having no formal involvement. Here is Carlile's description of the machine as it appeared in The Republican:

> Perhaps it will amuse you to be informed that in the new Temple of Reason my publications are sold by Clockwork!! In the shop is the dial on which is written every publication for sale: the purchaser enters and turns the hand of the dial to the publication he wants, when, on depositing his money, the publication drops down before him.[110]

The Crown, however, was certainly not amused. Use of the device was ineffective and both Carlile and his employee were convicted of selling blasphemous literature through the device.[111] Our society is full of these kinds of devices. The provider is usually very simple to identify: the city for parking meters, the selling company for soft drinks on the street or hotels. But cigarette dispensers are somewhat more difficult. Is the other party the shop-owner or the cigarette company? Although we do not know for sure we do not mind and just proceed with the transaction. In this respect, it is all about trust and credibility.

Pagallo, citing Chopra and White, also explained that, from the point of view of legal trust and credibility, for the acceptance of legal actions with legal effect, it must be clear on what mandate and on what legal attribution the agent is functioning.[112] For a vending machine, this is clear. For natural persons and AI entities it is not always clear. For natural persons representing legal persons we have to look up in official registers what their legal status in attribution of legal capacity encompass. If we make the comparison with the position of the Roman slave, it must also be considered that

---

[108] Voulon (2010).

[109] Raskin (2017), p. 10 [citing Segrave (2002)].

[110] Raskin (2017), p. 10–11 [referring to Carlile (1822)].

[111] See Footnote 110.

[112] Chopra and White (2011), p. 130, correctly remark, "to apply the respondent superior doctrine to a particular situation would require the artificial agent in question to be one that has been understood by virtue of its responsibilities and its interactions with third parties as acting as a legal agent for its principal."; Pagallo (2013), p. 132.

the relation between the slave and their master and the relation between the slave and society as a whole was more than instrumental. The slaves could perform a legal representative position, independent legal transactions and could appear as a witness in court. Moreover, the slave could be declared a "free man" by their master (manumission). This was not strange because at that time, on a population of one million people in Rome, there were 400,000 slaves. The position of the slave may be similar to the position of the robot a future society although declaring them "free men" as in the Millennium Man might be a step too far. Maybe robots could also hold peculium in the sense of a 'financial resource to be used without human control'. It is particularly crucial to determine to what extent it is desirable that robots will perform legal acts. Regarding a vacuum cleaner that position is clear. More complicated is the above-mentioned example of a social robot that buys groceries for a needy person or will order and then decide when and which medications should be administered. To hold a robot liable will only be efficient if the act cannot be tracked back to the original actor or "master" and to see what legal capacity this robot is performing a task, just as a representative of a legal person. In that case, and maybe other cases when it is not completely clear, obligatory insurance, financed by a general fund could provide a solution as also proposed in the EP Motion.[113]

9.3 Legal Acts

Why is it so important to define the shape of a certain legal personality for robots? If the robot acts with the intention to change the legal circumstances, be it autonomous and sentient, be it instrumental as instructed by another legal or natural person, they must also have a certain legal status beyond that of a legal object. In addition, we will need to find some form of liability that will ultimately best suit the practical qualifications and role of the robot in society. It must be deemed likely that robots in the surveillance and security areas as well as in the advisory and in the health sector, as well as in more exotic services, will play an important role without direct control by natural persons. The acts have to be recognized by other legal subjects based on trust and acceptance.

The responsibility of persons who are performing legal acts for others will ultimately rest with legal persons, a group or single identifiable individuals, the government, the official, political leaders and representatives accredited to a natural person. With the use of robots in those areas, that same responsibility will usually be traced to the same group and the robot will play a preparatory policy role or even a representative role.

[113]An obligatory insurance scheme, which could be based on the obligation of the producer to take out insurance for the autonomous robots it produces, should be established. The insurance system should be supplemented by a fund in order to ensure that damages can be compensated for in cases where no insurance cover exists. See European Parliament Report with recommendations to the Commission on Civil Law Rules on Robotics (2015/2103(INL)), RR\1115573EN.docx, p. 20.

It is conceivable that the robot will also be given a certain mandate attributed to them by authorities in the public sector to perform certain specified duties. Responsibility has to be determined. The arrest of a suspect by a "Robocop" has also to be secured legally. Legal and natural persons may be represented by robots in the future. This is a different situation than the legal representation by natural persons. This is only possible when it is established which specific competencies are relevant to the performance of the task of the robot. The attribution of competences has to be recognized by law. Only then there will be a legally credible acceptance of the legal effect of the performed acts by the robot.

The actions of an automated system may have legal implications. The advanced search robot meets other bots and will exchange some codes which can result in an agreement to reserve a seat or buy a product or service. The robot will enter a possible electronic agreement to be accepted by both electronic "parties" without any intervention or even confirmation by a natural person. Can this "Crawler Bot" still be considered an object if it has a kind of legal subjectivity?[114] This requires a clear explanation of the legal circumstances, preferably in the law and the contract, general terms and conditions.

Up until today, the fact that individual machines and devices were used for a purpose made the question of legal personhood irrelevant. A surgeon using a knife to make an incision in a patient and making a mistake cannot blame the knife or the knife producer for the mistake made by the surgeon (except in the case of a material error). In times of war, the producer of weapons cannot be held responsible for the casualties resulting from the war. However, the supreme commander, but also subordinates, may well be held responsible for possible war crimes. But what happens when these weapons are no longer instructed or directed by individuals? Or, if they provide information that will determine their operation without human intervention? If a drone is designed to recognize impending danger and subsequently destroys this danger without further instructions or intervention of individuals? For now, the destruction takes place by the action of a natural person using a joystick but even in that case the decision is based on data and intelligence that is going beyond the user. Several times, warnings were issued by concerned scholars and captains of industry concerning the dangers of autonomous AI weapons—so-called "killer robots"—recently in an open letter by the Future of Life Institute to the UN Convention on Certain Conventional Weapons.[115]

What is the qualification of the above case if the surgeon does not perform the surgery, but has recourse to sophisticated data supplied by a laser instrument that includes all medical information, including patient documentation? Or, if the computer or the social robot determines which drugs a patient requires, based on the patient records in the database? Why should an independent AI system not be capa-

[114] See European Parliament Report with recommendations to the Commission on Civil Law Rules on Robotics (2015/2103(INL)), RR\1115573EN.docx, p. 20.

[115] Future of Life Institute (2017) An Open Letter to The United Nations Convention On Certain Conventional Weapons. Available at: https://futureoflife.org/autonomous-weapons-open-letter-2017/. Accessed 21 August 2017.

ble to declare a valid death certificate. This should be an accepted legal act with legal consequences. Is there a distinction between an independently operating electronic system as an autonomous player and the use of this system as a tool? After all, in both cases the systems perform activities that have legal consequences.

Legal acts will be performed by persons, being legal entities. Automated systems, electronically or otherwise, are increasingly used in all kinds of relationships within our global society. Algorithms command the trading of the stock-market and buy and sell within milliseconds. The fact that these systems, robots and other devices can act independently and will create changes in legal relations will eventually have an effect on the position of legal persons, parties or third parties.

What is, ultimately, the difference between the agent in human form, the natural person and the robot representative?

In the command phase of the natural person or group of persons, the identification of the responsible player(s) normally is no problem. The difference in functional execution is not relevant. The use of search engines for finding tickets, drones for delivering packets to a client or sending of missiles on a perceived enemy will make no legal difference.

If the practical and legal responsibility can be traced back to a legal identification there is no change necessary in the legal position of the practical actor. The robot or AI system remains an instrument and legal object for which the legal entity remains responsible. Additionally, this includes the arrangements with respect to product liability in the case of a defective product.

For this aspect I refer to the exception in Article 185 sub paragraph e of Book 6 of the Dutch Civil Code where it is stated that a party who brings a product to the market of which, at the state of scientific and technical knowledge at the time he put the product into circulation, it was not possible to discover the existence of the defect of the product, will not be responsible for the defect.

And yet, this exception is pushed aside by producers of autonomous or semi-autonomous functioning cars like the Google car, Volvo and Tesla. It would also not be wise, at this time and from a public relations standpoint, if this risk would not be accepted by the producers. Regarding liability, a construction of risk liability and payment of damages from a kind of public foundation could be a solution and removing the "chilling effect" for further development of AI generated robots.

Even in the case of natural persons, as an attributed representative who loses their reason and sanity, the proceedings may be annulled as a non-deliberate disturbance of the system. One can draw a parallel with the robot in the latter cases; it can reduce the liability of the initiating individual in the use of this system or can exculpate all parties of the legal action, maybe even the robot itself, if the robot has legal responsibility.

This view I share with Voulon, in the sense that any legal effect which is caused by an autonomous and less autonomous system must be attributed to the natural or legal person who has made the decision to commission the system in its service operations.[116] This reasoning is based upon the functioning of electronic agents, described as:

> A computer program, or electronic or other automated means used independently to initiate an action, or respond to electronic messages or performances, on the person's behalf without review or action by an individual at the time of the action or response to the message or performance.[117]

One would apply the level of liability of the person or entity related to the degree of control exercised over the autonomous system, thereby also taking the legal effect into account. However, this would only be the case with regard to liability and accountability to the natural or legal person. The malfunction or failure of the autonomic system can be significant with regard to the recognition of the actor's legal liability. The autonomous system itself, however, can never bear any legal responsibility until there is a degree of legal personality and a certain acceptance of a legal position to perform legal actions with legal effect. A public register where the scope of legal competence of this entity is to be consulted would be a solution to enhance credibility.

Moreover, it would be helpful, in order to find a solution for this omission, to draw a parallel with the liability regulations as arranged in international regulations for electronic agents: The Uniform Electronic Transaction Act (UETA), the Uniform Computer Information Transaction Act (Ucita) and the Electronic Signatures Act (ESign). This could provide a model legal framework for autonomous entities to close agreements in a legally acceptable manner.

Ugo Pagallo presented the logical connection to existing forms of legal personhood for AI entities depending on their position and function, be it that more precise specifications of robot and their tasks can result in more specified legal subjectivity and legal competence:

"Independent legal personhood to robots with rights and duties of their own;
Some rights of constitutional personhood, such as those granted to minors and people with severe psychological illnesses, i.e., personhood without full legal capacity;
Dependent, rather than independent, personhood as it occurs with artificial legal persons such as corporations; and,
Stricter forms of personhood in the civil law field, such as the accountability of (some types of) robots for both contractual and extra-contractual obligations."[118]

---

116 Voulon (2010), concluding his dissertation.

117 Section 102 (a) (27) Uniform Computer Information Transaction Act (UCITA).

118 Going back to Teubner's analysis in the Rights of Non-Humans, the entry of new actors on the legal scene concerns all the nuances of legal agenthood, such as "distinctions between different graduations of legal subjectivity, between mere interests, partial rights and full-fledged rights, between limited and full capacity for action, between agency, representation and trust, between individual, group, corporate and other forms of collective responsibility." Pagallo (2013), p. 153 [referring to Teubner (2007)].

As Ugo Pagallo concludes in another book concerning contracting capability: "artificial agents should be able to qualify for independent legal personality" based on the task they have to perform.[119]

10 Conclusion and Steps into the Future

Although an autonomous system or robot, even with an independent intelligence and emotional capacity to function in our society, would not need to have a legal status that is similar to the rights and obligations of natural and legal persons in the positive law, change is imminent. The contours have to be defined. Even as an autonomous system passes the Turing test, this would not create any legal responsibilities per se. It is however advisable that certain forms of acting by autonomously functioning intelligent systems, such as social robots or legal enforcement robots, may be conceivable to obtain a certain form of attributed legal personhood to carry out their tasks. This is based on the essential requirement that there is a social and legal necessity justifying such an attribution.

The legal positioning of robots could be selected for an amendment of the law or possibly even a sui generis standard for certain autonomous robots. This legal positioning will be dependent on the degree of autonomy and social need. For the qualification of the robots, the grading of the ISO standards can serve as an example.[120] In the International Standardization Organization already a development can be seen to treat the role of the robot differently (in security) and to accept a standard for robot/human collaboration.[121]

One might also imagine that certain changes are made to the existing law in order to create a practical system representation of autonomous systems for the initial legal actor, the natural or legal person. These changes in the law will depend on a correct description of the reliability and trust of the representation by the robot, the purpose of the actions and the legal consensus of the legal entities involved. If these concepts are agreed upon, it will then be necessary to obtain the acceptance by the government and parliament to create or adapt a legal framework. As to how difficult and time consuming this process will be, reference can be made to the acceptance of the non-natural person in the positive law. The comparison with the rational, responsible actor as presented by Naffine probably will result in too many problems but certainly elements of this reasoning could be of help.

Currently, many AI systems are very difficult for users to understand. This is also increasingly true for those who develop the systems. In particular, neural networks

---

[119]Hildebrandt and Gaakeer (2013), p. 60.

[120]See, e.g., ISO 13,482: 2014 Specifies requirements and guidelines for the inherently safe design, protective measures, and information for use or personal care robots, in particular the following three types of personal care robots: mobile robot servant; physical assistant robot; person carrier robot.

[121]Human and robot system interaction in industrial settings is now possible thanks to ISO/TS 15,066, a new ISO technical specification for collaborative robot system safety.

are often "black boxes," in which the (decision-making) processes taking place can no longer be understood and for which there are no explanatory mechanisms.[122] This could necessitate a legal requirement to create a form of transparency as to how the systems work, to enhance trust and credibility of the acts leading to legal effect as also proposed in the EP motion on civil law rules on robotics.

AI and autonomous robots will be part of our future society. Integration of AI inside the human body will also occur. Our physical and informational integrity will be invaded, with or without our knowledge or consent. We already share a substantial part of our personal data with third parties and appear not really concerned by it. On top of that, governments and industries are forcing us to share even more personal information to regulate or protect the social system or to lower risks and costs of services and products.

The European General Data Protection Regulation describes the protection of personal data during processing in outdated terminology concerning AI.[123] Due to the non-technological orientation and the hinge on conventional directions of thinking, it hard to consider the GDPR sufficient to protect personal data in the age of AI.

Informational rights for the data subject and transparency of the process cannot be applied to integrated AI, certainly not if this is integrated into the physical functions of the human being. There is a significant risk of chilling effects for the development of AI and robotics if the GDPR has to be enforced on all AI applications.

In a report of the Science and Technology Committee of the UK Parliament, the need for unhindered but controlled applications of AI technology is stressed:

> It is important to ensure that AI technology is operating as intended and that unwanted, or unpredictable, behaviors are not produced, either by accident or maliciously. Methods are therefore required to verify that the system is functioning correctly. According to the Association for the Advancement of Artificial Intelligence: it is critical that one should be able to prove, test, measure and validate the reliability, performance, safety and ethical compliance – both logically and statistically/probabilistically – of such robotics and artificial intelligence systems before they are deployed.[124]

---

[122] Hildebrandt and Gaakeer (2013), p. 7.

[123] Regulation (EU) 2016/679.

[124] Interesting is the concluding recommendation of the Science and Technology Committee: "73. We recommend that a standing Commission on Artificial Intelligence be established, based at the Alan Turing Institute, to examine the social, ethical and legal implications of recent and potential developments in AI. It should focus on establishing principles to govern the development and application of AI techniques, as well as advising the Government of any regulation required on limits to its progression. It will need to be closely coordinated with the work of the Council of Data Ethics which the Government is currently setting up following the recommendation made in our Big Data Dilemma report.

74. Membership of the Commission should be broad and include those with expertise in law, social science and philosophy, as well as computer scientists, natural scientists, mathematicians and engineers. Members drawn from industry, NGOs and the public, should also be included and a programme of wide-ranging public dialogue instituted." Available at: https://publications.parliament.uk/pa/cm201617/cmselect/cmsctech/145/14506.htm#_idTextAnchor014. Accessed 25 October 2017.

For this reason, it will be necessary to develop some form of certification to determine whether the autonomously functioning robot can be accepted to process data of third parties and perform acts with legal capacity. Which interaction would be considered acceptable between parties will vary, depending on the function and of course the requirements of technological measures of protection of the robot as described above.

It is essential that we, as people, maintain control of the system as long as this has an added value. We would not want to be confronted with autonomous systems, which use the collection of all kinds of personal information and other available data for their own purposes. But, on the other hand, AI technology can only develop without chilling effects if it is commercially admitted to the consumer's daily life without too much legal constraint. The existence of a sui generis structure, comparable with the case of the artificial legal person in corporate law, may provide a solution. The Naffine definition of the Cheshire Cat combined with a Rational Actor model can form a rational basis for a legal framework comparable with the existing position of artificial legal persons.

At least, the following requirements of the AI entity have to be fulfilled to acquire a sui generis legal personhood:

Necessity in the 'human' society, socio-economic relevance, need for legal certification;
Determination of autonomous intelligence, Turing test like, 'human impression' level;
Sufficient social intelligence; The AI entity must be able to understand the socio-emotional and moral value of statements by other parties to respond appropriately so that there is an equivalent basis for consensus;
Being able to respond to changing circumstances; this aspect I would call 'adaptive or dynamic' intelligence;
Acceptance by other legal persons by creating trust and reliance for other legal and natural persons to integrate in economic, social and legal interactions;
A public register that specifies which robots will have specific legal competences for specified roles and tasks.

On top of this, an ethical code has to be developed on the basis of the EP motion that should also consider the use of different categories of robots, as well as the default rules needed for developers and producers of robotics.[125]

We are better off using our electronic, or better, technology-based servants to help us with the practical performance of our duties. The more intelligent the system is, all the more reliable the functionality will be. Give the robot a place in our legal system, maybe even with a form of digital peculium as proposed by Pagallo, giving them a limited resource that could also be used as a guarantee for possible mistakes or damages and opens the possibility of accountability for their autonomous acts.

[125]The proposed code of ethical conduct in the field of robotics will lay the groundwork for the identification, oversight and compliance with fundamental ethical principles from the design and development phase. EP motion, PE582.443v03-00, p. 21.

In a more extensive elaboration of this idea, one could establish a fund financed by a certain percentage of the earnings by robots to guarantee any losses or damages. Though it will have to be a select group of AI entities that qualify for a new form of legal personhood and economic personality. In that respect, the robot will be active in the social and economic functioning of society. This can also concern the public sector. A certain trust in the acts of robots and recognition of their identity will prove to be essential.

But we have to keep in mind that we still have to control the developments and not end up with the rather pessimistic "post-human" idea described by Yuval Noah Harari in his famous book Homo Deus. In this account, science will move in the direction that all organisms are algorithms, life is data-processing, intelligence will be separated from consciousness and the hyper-intelligent algorithms will know us better than we know ourselves.[126] This means that super-intelligent algorithms will decide how our life, or whatever existence will be left, will develop without any human influence.

References

Asimov, I. (1976). *The bicentennial man and other stories*. London: Victor Gollancz Ltd.

Asimov, I., & Silverberg, R. (1993). *The positronic man*. London: Doubleday.

Berriat Saint-Prix, J. (1829). *Rapport et recherches sur les procès et jugemens relatifs aux animaux*. Paris: Imprimerie de Selligue.

Bertolini, A. (2013). Robots as products: The case for a realistic analysis of robotic applications and liability rules. *Law, Innovation and Technology, 5*(2), 214–227.

Bodin, J. (1955). *Les six Livres de la Republique* (trans: M. J. Tooley). Oxford: Blackwell.

Bostrom, N. (2014). *Superintelligence: Paths, dangers, Strategies*. Oxford: Oxford University Press.

Brownlie, I. (1990). *Principles of public international law*. Oxford: Clarendon Press.

Bryson, J. J., Diamantis, M. E., & Grant, T. D. (2017). Of, or, and by the people: The legal lacuna of synthetic persons. *Artificial Intelligence Law, 25*, 273–291.

Burkhart, L. (2016). Symposium—Governance of emerging technologies: Law, policy, and ethics. *Jurimetrics, 56*, 219–222. Available at: https://www.americanbar.org/content/dam/aba/administrative/science_technology/2016/governance_in_emerging_technologies.authcheckdam.pdf. Accessed September 12, 2017.

Carlile, R. (1822). To the republicans of the island of Great Britain. *Republican, 16*(V).

Chopra, S., & White, L. F. (2011). *A legal theory for autonomous artificial agents*. Ann Arbor: The University of Michigan Press.

Clarke, A. C. (1973). *Profiles of the future: An inquiry into the limits of the possible*. New York: Harper & Row.

Crawford, J. R. (2012). *Brownlie's principles of public international law* (8th ed.). Oxford: Oxford University Press.

Darling, K. (2016). *Electronic love, trust, & abuse: Social aspects of robotics*. Workshop at the University of Miami, Conference "We Robot", April 2016.

Delvaux, M. (2017). *Report PE582.443v01-00 with recommendations to the Commission on Civil Law Rules on Robotics* (2015/2103(INL)) http://www.europarl.europa.eu/sides/getDoc.do?pubRef=-//EP//NONSGML+REPORT+A8-2017-0005+0+DOC+PDF+V0//EN. Accessed December 8, 2017.

[126] Harari (2017), last sentences.

Descartes, R. (1973). *Principia philosophiae*. Paris: Vrin.
Dewey, J. (1926). The historic background of corporate legal personality. *Yale Law Review, 35*(6), 655–673.
Erven D onder de Linden en zoon. (1831). Boekzaal der geleerde wereld: En tijdschrift voor de Protestantsche kerken in het koningrijk der Nederlanden, Amsterdam.
Fagundes, D. (2001). What we talk about when we talk about persons: The language of a legal fiction. *Harvard Law Review, 114*(6), 1745–1768.
Future of Life Institute. (2017). *An open letter to the United Nations convention on certain conventional weapons*. Available at: https://futureoflife.org/autonomous-weapons-open-letter-2017/. Accessed August 21, 2017.
Gardner, H. (1993). *The theory of multiple intelligences*. New York: Basic Books.
Geldart, W. M. (1911). Legal personality. *Law Quarterly Review, 27*, 90–108.
Giliker, P. (2011). Vicarious liability or liability for the acts of others in tort: A comparative perspective. *Journal of European Tort Law, 2*(1), 31–56.
Good, I. J. (1965). Speculations concerning the first ultraintelligent machine. In F. L. Alt & M. Rubinoff (Eds.), *Advances in computers* (Vol. 6, pp. 31–88). Cambridge: Academic Press.
Harari, Y. N. (2017). *Homo deus: A brief history of tomorrow*. New York: Random House.
Hildebrandt, M., & Gaakeer, J. (Eds.). (2013). *Human law and computer law: Comparative perspectives*. Dordrecht: Springer.
Hobbes, T. (1651). Chapter xvi: Of persons, authors, and things personated. In *Leviathan*. London: Andrew Crooke.
Hutchinson, A. (2014). The Whanganui river as a legal person. *Alternative Law Journal, 39*(3), 179–182.
Kurzweil, R. (1990). *The age of intelligent machines*. Cambridge: The MIT Press.
Lovejoy, A. O. (1936). *The great chain of being: A study of the history of an idea*. Cambridge: Harvard University Press.
Mayer, C. J. (1990). Personalizing the impersonal: Corporations and the bill of rights. *Hastings Law Journal, 41*(3), 577–667.
Miller, C. A., & Bennett, I. (2008). Thinking longer term about technology: Is there value in science fiction-inspired approaches to constructing futures? *Science and Public Policy, 35*(8), 597–606.
Minors, D. (2017). *Can you read my mind?* Available at: https://www.wits.ac.za/news/latest-news/research-news/2017/2017-09/can-you-read-my-mind. Accessed October 11, 2017.
Mori, M. (2012). *The uncanny valley: The original essay by Masahiro Mori*. https://spectrum.ieee.org/automaton/robotics/humanoids/the-uncanny-valley. Accessed October 15, 2017.
Naffine, N. (2003). Who are law's persons? From Cheshire cats to responsible subjects. *Modern Law Review, 66*(3), 346–367.
Ohlin, J. D. (2005). Is the concept of person necessary for human rights? *Columbia Law Review, 105*, 209–249.
Pagallo, U. (2013). *The laws of robots: Crimes, contracts, and torts*. Dordrecht: Springer.
Raskin, M. (2017). The law and legality of smart contracts. *Georgetown Law Technology Review, 304*(1). https://www.georgetownlawtechreview.org/the-law-and-legality-of-smart-contracts/GLTR-04-2017/. Accessed October 20, 2017.
Richards, N. M., & King, J. H. (2014). Big data ethics. *Wake Forest Law Review, 49*, 393–432.
Russell, S., & Norvig, P. (2010). *Artificial intelligence: A modern approach* (3rd ed.). New Jersey: Pearson Education.
Safi, M. (2017). *Ganges and Yamuna rivers granted same legal rights as human beings*. https://www.theguardian.com/world/2017/mar/21/ganges-and-yamuna-rivers-granted-same-legal-rights-as-human-beings. Accessed May 13, 2017.
Schaerer, E., Kelley, R., & Nicolescu, M. (2009). Robots as animals: A framework for liability and responsibility in human-robot interaction. In Robot and human interaction communication. RO-MAN 2009—The 18th IEEE international symposium on robot and human interactive communication. *Journal Advanced Robotics, 24*(13), 1861–1871.

Science and Technology Committee. (2016). *Robotics and artificial intelligence*. https://publications.parliament.uk/pa/cm201617/cmselect/cmsctech/145/14506.htm#_idTextAnchor014. Accessed October 25, 2017.
Segrave, K. (2002). *Vending machines: An American social history*. Chicago: McFarland & Co.
Shoyama, R. (2005). Intelligent agents: Authors, makers, and owners of computer-generated works in Canadian copyright law. *Canadian Journal of Law and Technology, 4*(2), 129–140.
Solum, L. B. (1992). Legal personhood for artificial intelligences. *North Carolina Law Review, 70*(4), 1238–1239.
Teubner, G. (2007). *Rights of Non-humans? Electronic agents and animals as new actors in politics and law*. Badia Fiesolana: European University Institute.
The Global Slavery Index. (2016). https://www.globalslaveryindex.org/findings/. Accessed October 12, 2017.
Tjong Tjin Tai, T. F. E. (2016). Private law for homo digitalis, use and maintenance. In *Preliminary advice for NVJ* (p. 248).
Turing, A. M. (1950). Computing machinery and intelligence. *Mind, New Series, 59*(236), 433–460.
Vinge, V. (1993). The coming technological singularity: How to survive in the post-human era. *NASA, Lewis Research Center, Vision, 21*, 11–22. Available at: https://edoras.sdsu.edu/~vinge/misc/singularity.html. Accessed October 25, 2017.
Von Bar, C., & Clive, E. (Eds.). (2009). *Principles, definitions and model rules of European private law: Draft Common Frame of Reference (CDFR)*. Munich: Sellier. European Law Publishers GmbH.
Voulon, M. B. (2010). Automatisch Contracteren. *Dissertation*, Leiden University.
Weschler, D. (1955). *The range of human capacities*. Baltimore: Williams & Wilkins.
Wiener, N. (1950). *The human use of human beings*. Houghton Mifflin: Eyre & Spottiswoode.

The Peculiar Case of the Mushroom Picking Robot: Extra-contractual Liability in Robotics

Ioannis Revolidis and Alan Dahi

Abstract This chapter focuses on the extra-contractual liability of robots. It shows that robot-specific difficulties facing the legal system can be found in other areas of the law, and that the law has successfully addressed the difficulties. As such, a specific "(Liability) Law of the Robot" is not needed. Moreover, robots are too diverse a category to permit a uniform approach of dealing with the liability of their acts. Robots, and the underlying Artificial Intelligence, will need to be assessed against their purposes and capabilities, respectively. The contribution does not intend to offer a detailed answer on how exactly the problem of extra-contractual liability of robots shall be addressed, considering that such a discussion goes beyond a book chapter. It represents a first effort to explore the methodological particularities of the problem. It will, therefore, only include detailed insights to the extent necessary for the relevant methodological discussion. Without any intention of oversimplifying the problem of the civil accountability of robots, as the detailed nuances of the probable solutions definitely need further refinement, the chapter assumes that the traditional risk distribution mechanisms of civil liability systems can provide for a solid framework that can be processed further in order to adequately meet the particularities of robots. Drawing from the "Law of the Horse" debate, the chapter neither pleads for technological insensitivity nor does it proclaim that technological utopianism shall be the method to replace it, but it suggests that the lessons from regulating the Internet might point to a creative synthesis of technological advancements and traditional regulatory mechanisms, so that both are represented equally in the new set of rules that is meant to regulate new and disruptive phenomena, such as the social and economic impact of robots.

Keywords Robots · Civil liability · Torts · Agency · Legal personhood

---

I. Revolidis (✉)
Institute for Legal Informatics, Leibniz Universität Hannover, Hannover, Germany
e-mail: ioannis.revolidis@iri.uni-hannover.de

A. Dahi
Grand Cayman, Cayman Islands

M. Corrales et al. (eds.), *Robotics, AI and the Future of Law*, Perspectives in Law, Business and Innovation, https://doi.org/10.1007/978-981-13-2874-9_3

1 Introduction

In 2017, the European Parliament issued a resolution with recommendations requesting the European Commission to submit a proposal for a directive on civil law rules for robotics (Parliament's Resolution),[1] by all appearances paving the way for a specialized "Law of the ~~Horse~~[2] Robot."

The Parliament's Resolution displays a sweeping grandeur. It references in its first paragraph Mary Shelley's "Frankenstein," the Greek tale of Pygmalion,[3] the Golem of Prague,[4] and also the 1920 play "R.U.R."[5] by Karel Čapek.[6]

In its second paragraph, the resolution boldly states that because:

> humankind stands on the threshold of an era when ever more sophisticated robots, bots, androids and other manifestations of artificial intelligence ('AI') seem to be poised to unleash a new industrial revolution, which is likely to leave no stratum of society untouched, it is vitally important for the legislature to consider its legal and ethical implications and effects, without stifling innovation.[7]

Specifically, with regard to liability, the resolution submits that in certain scenarios the traditional rules are insufficient, such as where a robot can learn and take autonomous decisions, and where machines directly conclude contracts and implement them.[8]

This chapter seeks to understand, specifically with regard to extra-contractual liability, whether traditional rules are truly insufficient—and if so, in which circumstances that may be.[9] We will begin by defining artificial intelligence and robotics (Sect. 2). We will then pick up on Calo's seminal investigation "Robotics and the Lessons of Cyberlaw"[10] to present—and critique—some of the considerations why robotics might indeed demand a certain re-evaluation of the legal status quo (Sect. 3). Subsequently, we will look at existing metaphors for robots in an attempt to understand how the law might deal with the questions that robots raise (Sect. 4). Based on this analysis, we will propose which of the actors could be held liable (Sect. 5) as a matter of public policy before summarizing the results (Sect. 6).

---

[1]European Parliament (2017).

[2]See Easterbook (1996).

[3]Pygmalion was a sculptor who fell in love with one of his statues, which came to life after Aphrodite granted the sculptor's wish for this.

[4]The Golem of Prague is the tale of a golem created out of clay and brought to life to protect the Jewish ghettos in 16th century Prague.

[5]Čapek's robots were actually created out of synthetic organic matter; like their machine counterparts in the Terminator movie series, they eventually rebelled against their human lords. R.U.R. also introduced the word 'robot' to the world, a play on the Czech "robota," which means a type of forced labor.

[6]European Parliament (2017), Recital A.

[7]European Parliament (2017), Recital B.

[8]European Parliament (2017), Recitals AA-AI.

[9]Jurisdictional questions, such as may arise from tele-operated robots, are excluded.

[10]Calo (2015).

2 Delineating Artificial Intelligence and Robotics

2.1 Artificial Intelligence Defined

Artificial Intelligence (AI) can be broadly split into three types: Artificial Narrow Intelligence (ANI), Artificial General Intelligence (AGI), and Artificial Super Intelligence (ASI).[11]

Today's AI deployment falls under the category of ANI, which can be defined as "the ability of machines to resemble human capabilities in narrow domains."[12] The levels of ability may be quite different; one need only compare IBM's *Watson* competing on and wining *Jeopardy*,[13] a popular television quiz show, with an intelligent chatbot like *Woebot*,[14] which offers science-backed psychological support.

AGI, in contrast, is the holy grail of current AI research. An AGI would have human capabilities across a number of domains.[15] As such, it would be a true human-equivalent AI.

ASI is a whole different category altogether. It is an intelligence that is "much smarter than the best human brains in practically every field, including scientific creativity, general wisdom and social skills."[16] This is the level of AI that worries people, such as the philosopher and futurist Nick Bostrom, Microsoft's Bill Gates, or Tesla Motor's and SpaceX's Elon Musk.[17] As Tim Urban puts it:

> If our meager brains were able to invent wifi, then something 100 or 1,000 or 1 billion times smarter than we are should have no problem controlling the positioning of each and every atom in the world in any way it likes, at any time – everything we consider magic, every power we imagine a supreme God to have will be as mundane an activity for the ASI as flipping on a light switch is for us. Creating the technology to reverse human aging, curing disease and hunger and even mortality, reprogramming the weather to protect the future of life on Earth – all suddenly possible. Also possible is the immediate end of all life on Earth. As far as we're concerned, if an ASI comes to being, there is now an omnipotent God on Earth – and the all-important question for us is: Will it be a nice God?[18]

Once AGI has been achieved, and definitely with ASI, the question will arise whether the AI has also achieved consciousness, or self-awareness, leading us to both age-old and modern discussions into what it means to be "human," to have free-will, and to deserve complete, constitutional personhood. Artificial intelligence will become artificial life. These specific issues are beyond the scope of this chapter, however, as is generally ASI with its unforeseeable impact on society.

---

[11] Lords Committee (2018), p. 15, para. 14 et seq; Privacy International, Article 19 (2018), p. 6; Urban (2015).

[12] Privacy International, Article 19 (2018), p. 6.

[13] Markoff (2011).

[14] Fitzpatrick et al. (2017) and also https://woebot.io/.

[15] Privacy International, Article 19 (2018), p. 6.

[16] Bostrom (1998).

[17] See Gibbs (2015) for an overview.

[18] Urban (2015).

2.2 Robots Defined

Robots, in a sense, are the physical embodiment of AI. Being purely algorithmic, AI can generate information, but the AI itself cannot, directly, physically affect its environment.[19] It is a mere ghost without a shell. However, once integrated into a machine, the ghost gains a physical shell with which it can affect change in its environment by sensing, "thinking," and then acting.[20] It becomes "embodied."[21] Consequently, a robot as defined in this chapter is the result of a merging of AI and machine so that the AI can independently and directly act on the physical world.[22]

The sense-think-act definition of robots for the purposes of legal investigation is a common one.[23] The Parliament's Resolution, too, recommends establishing a common definition and classification of "smart robots," based on the sense-think-act characteristics of:

i. the capacity to acquire autonomy through sensors and/or by exchanging data with its environment (inter-connectivity) and the analysis of those data;
ii. the capacity to learn through experience and interaction;
iii. the form of the robot's physical support;
iv. the capacity to adapt its behavior and actions to the environment.[24]

The Parliament's definition also captures the essence of what "think" actually means. It is the capacity to learn, such that a robot's behavior cannot be determined in advance because the outcome will depend on the outcome of the robot's "thought" process. A robot that follows the rule "turn left when forward motion is blocked" may sense the obstruction and act on it, but it does not think about what it sensed. It is pre-programmed. Thinking requires that the robot is capable of assessing what it sensed in order to decide how to act.

While it may be argued that not being able to determine in advance how the robot will act is merely a reflection of our own limited understanding of how AI works, similarly to how neuroscience is currently unable to resolve the free will debate,[25] relying on whether a robot can show unexpected behavior ("Which of the multiple options will it choose?") is a sufficient and suitable test for autonomy.

---

[19] An example is an AI that is fed data from a health check and which then recommends certain health measures that a patient can follow. See also Calo (2015), p. 531.

[20] Calo (2015), p. 529.

[21] Calo (2015), p. 531; Urban (2015), who describes a robot as an AI brain with a body.

[22] See also Calo (2015), p. 531.

[23] See Footnote 20.

[24] European Parliament (2017), Annex.

[25] See, e.g., Saigle (2018).

3 Exceptional Robots—Lessons from Cyberlaw

According to Calo's seminal investigation into the lessons from cyberlaw for robotics, the reason why robots will, to a certain extent, challenge the current legal system is that they are an exceptional[26] technology, marked by the qualities of embodiment, emergence, and social valence.[27] We will take a closer look at each of these elements and their meaning for the law.

3.1 Embodiment

Embodiment is the consequence of a physical body being able to act on information and directly affect the world at large.

Calo puts forward that embodiment represents a challenge for the law because the intangible (i.e., information) generally does not give rise to actions under product liability.[28] For this, he cites inter alia the U.S. case of *Winter v. G.P. Putnam's Sons,*[29] in which the court held that two mushroom pickers who were poisoned as a result of relying on wrong information in an encyclopedia on mushrooms could not claim damages from the publisher based on product liability.[30] Product liability requires a defect in a tangible product. However, the physical product, i.e., the book, was not defective; its intangible content was.[31]

According to Calo, embodiment muddies the border between informing and acting.[32] The book in *Winter* merely informed. It did not act; it was the humans that picked and prepared the mushrooms.

While Calo's reference to *Winter* serves to illustrate the problems of embodied information, it is incomplete and thus gives a false impression. Embodied information can in some cases give rise to a product liability claim.

As the court in *Winter* states, aeronautical charts have been held to be products for the purpose of product liability law.[33] Ultimately, however, the court differentiates between aeronautical charts and books such as the encyclopedia by drawing an analogy between charts and a compass, explaining that both are tools that may be used to guide an individual where knowledge of natural features is required. In contrast, a book such as the mushroom encyclopedia is rather "like a book on how

---

[26] A technology is seen as exceptional for the law where it will not just require a superficial change to the legal system, such as a minor adjustment of laws, but where it challenges fundamental doctrines and thus invites their recalibration Calo (2015), p. 553.

[27] Calo (2015), p. 532.

[28] Calo (2015), p. 535 et seq.

[29] 938 F.2d 1033 (9th Cir. 1991).

[30] Calo (2015), p. 536.

[31] 938 F.2d 1033 (9th Cir. 1991), para. 5 et seq.

[32] See Footnote 30.

[33] 938 F.2d 1033 (9th Cir. 1991), para. 11 et seq., citing a number of decisions.

to use a compass or an aeronautical chart" and as such represents "pure thought and expression,"[34] i.e., the intangible.

We can imagine a mushroom picking robot with ANI capabilities. It identifies and picks edible mushrooms. Alas, it mistakenly picks a poisonous mushroom for the family it serves, who, after partaking of the tasty but ill-fated dinner, suffers severe discomfort. If a court likens an aeronautical chart to a tool for product liability purposes, it is not a stretch to believe that the same might occur for a robot that faultily picks a poisonous mushroom—at least regarding the embodiment of information.

3.2 Emergence (Vs. Autonomy)

The next factor Calo assesses is that of "emergence," i.e., the intelligent, higher-level "thinking" displayed by robots. He prefers the term to the more commonly used "autonomy," which he believes suggests a certain intent in actions—something (certainly ANI-level) robots do not have.[35]

The Parliament's Resolution defines a robot's autonomy as "the ability to take decisions and implement them in the outside world, independently of external control or influence; whereas this autonomy is of a purely technological nature and its degree depends on how sophisticated a robot's interaction with its environment has been designed to be."[36]

"Emergence" is used across a variety of fields, including philosophy, systems theory, and various sciences.[37] It is generally understood as the concept of the whole being more than the sum of its parts, where individuals act independently but within a self-organizing system that exhibits a higher-order intelligence, and where low-level rules lead to higher-level sophistication.[38] This concept also applies to AI because "intelligence emerges from the interaction of the components of the system," without there being a single component that is the source of the intelligence displayed.[39]

The extraordinary bridges built by army ant colonies are perfect examples of emergent higher-level behavior. An individual ant will (i) slow down when it reaches a gap it cannot cross, and (ii) freeze in place as soon as a still faster moving ant from behind walks over it.[40] The individual ant's behavior is very simple, but the resultant bridges of interlinked ant bodies are extraordinarily complex. Swarms of robots have been programmed to exhibit similar behavior.[41]

[34] 938 F.2d 1033 (9th Cir. 1991), para. 11.

[35] Calo (2015), p. 538 et seq. See also Balkin (2015), p. 51, who agrees that "emergence" is preferable to "autonomy," which raises difficult questions regarding the status of AI.

[36] European Parliament (2017), Recital AA.

[37] See Goldstein (1999).

[38] See, e.g., Johnson (2001), 29.4/571, Wolchover (2017).

[39] Brooks (1991), p. 16.

[40] Hartnett (2018b).

[41] Hartnett (2018a).

The ant army example reveals the difference between autonomy and emergence. The individual ant acts autonomously while the colony displays emergent, higher-level qualities. The importance of autonomy is, in a sense, even touched upon by the court in *Winter*, when it calls an aeronautical chart a highly technical tool.[42] A tool is something that is used; it is the object of an action, not the subject. The question is then: When does a robot pass the threshold from tool to autonomous action? Vladeck distinguishes between tools *used* by humans and tools (machines) *deployed* by humans.[43] Deployed tools can operate in circumstances unforeseen by the machine's creators precisely because they are autonomous. As will be discussed below, autonomy makes it difficult to ask to whom liability shall fall for harm caused by an autonomously acting robot.[44] Autonomy also leads us back to the sense-think-act paradigm mentioned as part of the definition of "smart robot."

Consequently, we believe that the autonomy of robots is more important from a legal perspective than emergence is. Emergence still plays a role, but in our opinion rather with regards to AI and robots being generative technologies,[45] with multiple layers of hardware, software, and protocols embedded within a social layer.[46]

3.3 Social Valence

The final element of Calo's trio is that of "social valence," i.e., the fact that robots elicit emotional responses from people in a social context. Put differently, we anthropomorphize robots.

As Calo explains:

> If contemporary psychology is struggling with how to categorize robotics given its liminal status between agent and object, it should not surprise us that criminal, tort, and other law may as well. Speaking very broadly, the law tends to assume a dichotomy between individuals and tools.[47]

By way of example, he mentions that in medical malpractice suits a common complaint is that the patient/doctor interaction was inadequate—a complaint is never that the patient was not given the opportunity to make her acquaintance with the scalpel used in the operating room.[48]

In which category shall a medical robot be placed? It is, in a sense, a tool no different than a scalpel, albeit a smart one. How patients, doctors, and others react to a robot compared to a scalpel, on the other hand, is very different.

---

[42] 938 F.2d 1033 (9th Cir. 1991), para. 12.

[43] Vladeck (2014), p. 121.

[44] See also Balkin (2015), p. 52; Scherer (2016), pp. 363 et seqq.

[45] Zittrain (2006).

[46] See also Balkin (2015), p. 53.

[47] Calo (2015), p. 546.

[48] Calo (2015), p. 547.

Balkin submits that anthropomorphizing robots indicates that "people are willing to substitute [robots] for animals or human beings in certain contexts and for certain purposes."[49] He calls this the "substitution effect,"[50] and explains how:

> The substitution of robot for living thing may be innocent, emotional, almost instinctual. The patient who blames a surgical robot for a botched procedure projects a partial humanity – and hence responsibility – onto the technology. The soldier who mourns the loss of his bomb-disarming robot projects onto the robot human qualities of comradeship, courage, and commitment to fellow soldiers. When a companion robot who operates in our home sends personal data about us to a corporation, we feel betrayed, when we would never think that a camera and a microphone could betray us.[51]

The consequence is that a layperson might "feel" that the law should hold the robot liable. And, as the rights of minorities show, the law frequently develops by societal demands.

3.4 An Exceptional Trio

Easterbrook famously declared in 1996 that for the same reasons why there is no need for a law of the horse there need also be no law of the Internet[52]: Any "course on 'The Law of the Horse' is doomed to be shallow and to miss unifying principles" across the law.

Is robotics a field of law that is exceptional and worthy of its own course, or will the legal issues robots raise be dealt with as something one may very well specialize in, but still as something that does not necessarily warrant a distinct field of study?

Based on the impact of embodiment, emergence/autonomy, and social valence, Calo believes robotics will prove to have an intermediate level of legal exceptionalism, stating that:

> A technology is not exceptional merely because it creates one or more small changes in the law, or because it reveals, at the margins, that an existing interpretation of a particular doctrine is incomplete. By the same token, a technology need not occasion a literal breakdown in the rule of law or prove the source of entirely novel doctrines to qualify. Rather, a technology is exceptional when its introduction into the mainstream requires a systematic change to the law or legal institutions in order to reproduce, or if necessary displace, an existing balance of values.[53]

---

[49] Balkin (2015), p. 57.

[50] See Footnote 49.

[51] Balkin (2015), p. 58.

[52] Easterbrook (1996), pp. 207 et seqq. Though lawyers do specialize in equine law.

[53] Calo (2015), p. 552.

Some of the areas that Calo believes will be so affected are that of *mens rea* in criminal law[54]; of administrative law as a result of needing new, specialized administrative agencies[55]; and of foreseeability in tort law.[56]

What strikes us, however, is that the trio of embodiment, autonomy, and social valence all belong to the standard operating model of biological life. A human, an octopus,[57] a donkey—even a slime mold[58]—all take in information and physically act on it. This gives us embodiment. They also all act sufficiently autonomously to pass the sense-think-act test. And they even trigger a social valence—albeit some representatives of the biological realm more than others, such as a human or a cuddly donkey compared to an insect or a slime mold.

If this is indeed the case, does it make sense to differentiate between digital and biological intelligence? Or might we be able to place robots somewhere within the law's treatment of their biological cousins regarding civil/extra-contractual liability? As Balkin puts it: Are robots "special-purpose animals" or "special-purpose human beings"?[59] These are the questions the next section attempts to answer.

4 The Extra-contractual Liability of the Robot

The Parliament's Resolution foresees that a robot's autonomy will cause difficulty in attributing liability for the robot's actions because:

> Whereas the more autonomous robots are, the less they can be considered to be simple tools in the hands of other actors (such as the manufacturer, the operator, the owner, the user, etc.); whereas this, in turn, questions whether the ordinary rules on liability are sufficient or whether it calls for new principles and rules to provide clarity on the legal liability of various actors concerning responsibility for the acts and omissions of robots where the cause cannot be traced back to a specific human actor and whether the acts or omissions of robots which have caused harm could have been avoided.[60]

Solving this problem is seen as critical by the Parliament, which suggests as solutions to investigate: a compulsory insurance scheme (such as currently exists for cars); a compensation fund for damages not covered by any insurance; the consideration of limited liability for the actors involved in the "making" of the robot as a benefit for contributing to a compensation fund; a robot registry with all relevant details on liability (similar to a companies registry); and even a specific legal sta-

[54]Calo (2015), p. 554.
[55]Calo (2015), p. 555 et seq.
[56]Calo (2015), p. 554 et seq.
[57]Judson (2017).
[58]Yong (2010).
[59]See Footnote 49.
[60]European Parliament (2017), Recital AB.

tus/personhood for certain categories of robots, which would elevate the robot to being able to be held directly liable.[61]

In order to understand whether the Parliament's suggestions are necessary, we will look at how the law handles extra-contractual liability where one autonomous being acts on behalf of or under the direction of another, namely regarding agents, minors and others under supervision, slaves, and animals. Firstly, some general principles of tort law. As tort law differs considerably from jurisdiction to jurisdiction,[62] we will use the Principles on European Tort Law (PETL) as a basis against which to assess the impact robots may have on tort law, though with some "peeks" at German law in order to anchor the principles to a concrete jurisdiction or where the PETL do not address the issue.

The PETL were drafted by the European Group on Tort Law (EGTL), a network of academics that used hypotheticals and country and comparative reports to identify principles common to both common law and civil law European jurisdictions.[63] The PETL are not a draft code,[64] and neither are they a restatement of the law of torts.[65] And while they do strive to state common principles, it should be noted that, where the national differences are too great, the PETL also contain proposals that go beyond a mere reflection of common principles.[66]

Liability requires the elements of: damage to a legally protected interest[67]; causation for the damage by an activity, which could be either an act or omission[68]; and a recognized base for the liability.[69] Causation and the base of liability are the relevant elements for the purposes of this section; damage will generally not be an issue.

Causation is given where an activity is *conditio sine qua non* for the damage, i.e., where the damage would not have occurred in the absence of the activity.[70] As this element is too encompassing, it is limited by requiring a base of liability. The bases are fault; an abnormally dangerous activity; and an auxiliary acting on behalf of the person deemed liable, or, more generally, being in charge of another.[71]

"Fault" is the intentional or negligent breach of the required standard of conduct[72]; this can also encompass product liability resulting from any "deviation from standards that are reasonably to be expected."[73] With regards to product liability, the burden of

[61] European Parliament (2017), paras 49 et seq.

[62] EGTL, Section I.

[63] EGTL, Section III.

[64] Koch (2007), p. 108 et seq.

[65] Koch (2007), p. 110.

[66] See Footnote 65.

[67] Article 2:101 PETL.

[68] Article 3:101 PETL.

[69] Title III PETL.

[70] Article 3:101 PETL.

[71] Articles 1:101, 6:101 PETL.

[72] Article 4:101 PETL.

[73] Article 4:202 PETL.

proof is reversed; the manufacturer must prove that the required standard of conduct was met to not be liable.[74]

An abnormally dangerous activity gives rise to strict liability; an activity is deemed abnormally dangerous where there is a foreseeable and highly significant risk of damage even when all due care is given, and where the activity is not a common one.[75]

Finally, being in charge of another, either because the individual is a minor, or because an auxiliary (agent) is engaged, are questions that will be investigated below in their own section.

Applied to our mushroom-picking robot, liability could be directed against the robot itself as the acting agent, though this would require legal personhood. It could also be directed against the manufacturer under product liability rules (notwithstanding the problem of determining who the manufacturer is in multi-component cases), and finally against the owner/supervisor of the robot.[76] To whom liability shall fall will be investigated below in Sect. 5.

4.1 *Robots and Agents*

Broadly speaking, an agent is someone who is used by another, the principal, to perform a task, the same as how a robot might be used to perform a task. The general principle for extra-contractual liability found in both the common law and civil law is that the principal will be held liable for the agent insofar as the latter was acting within their assigned scope and the agent violated the appropriate standard of contract.[77] This follows from the doctrine of *respondeat superior*, which assigns responsibility within a hierarchy to the superior.[78] An employee who unintentionally causes harm in the course of their employment activities, for example a waiter spilling wine over a customer's blouse, should generally not be the target of liability claims; the restaurant owner profits from the waiter's labor and the restaurant owner should, within reason, compensate for any damages incurred.

Robots will come in many different types with different purposes. Some will have rather narrow purposes, such as mowing the lawn, while some will be generalists, for example assisting much as a butler would, running errands that would range from being a chauffeur to doing the shopping and dropping off laundry at the dry-cleaners.[79]

The law differentiates between "detours" and "frolics" of agents—the former are necessary adaptations of an agent to fulfill the task where the circumstances have

[74] See Footnote 73.

[75] Article 5:101 PETL.

[76] See also Petit (2017), pp. 18 et seqq.

[77] Article 6:102 PETL; § 831 para 1 BGB; Asaro (2011), p. 178 et seq.

[78] Asaro (2011), p. 178.

[79] See Asaro (2011), p. 179, which inspired the example.

changes, the latter are private escapades on the principal's time. A detour will not change the default liability rule, while a frolic, being of a private nature, will directly cause the agent to be liable.[80]

The more autonomous and general-purpose a robot, the more difficult it will be to decide whether it was on a detour or a frolic. As Asaro points out,[81] though, if a robot is deemed to have been on a frolic, shall the robot itself be held liable? This is a question that will be investigated below in Sect. 5.

4.2 *Robots, Minors, and Other Persons Under Supervision*

Persons under the supervision of others are another category similar to that of an autonomous robot and its "supervisor," be that an owner or another person tasked with the duty of supervising it. As such, it warrants a closer look.

Under German tort and negligence law, the statutory or contractual supervisor of minors and of persons who, because of their physical or mental condition, require supervision, will be held liable under tort for any damages caused by those under the supervisor's supervision unless the supervision was properly conducted or the damage would have also occurred otherwise.[82] It should be noted that the "unless" is because the law reverses the burden of proof such that the supervisor must exculpate themselves by proving they fulfilled the standard of care required of them in the concrete situation.[83]

The reason for this default liability of the supervisor is that minors and persons placed under the supervision of another are deemed to be insufficiently capable of acting appropriately. They are autonomous, but they need a minder. This approach is fairly standard and is also reflected in the PETL.[84]

Translated to robots, the rule would apply where a robot is autonomous and can fulfill its tasks (think of a young child helping by putting its dinner plate into the dish washer) but where in certain environments its capabilities are challenged and it would need supervision (think of a young child helping at a friend's house by clearing a dinner table with fancy porcelain).

---

[80]See for the common law Asaro (2011), p. 179. Civil law would come to the same conclusion because the action that caused harm would either be determined to be within the assigned scope (a detour) or not (a frolic).

[81]Asaro (2011), p. 179.

[82]§ 832 paras 1, 2 BGB.

[83]§ 276 paras 1, 2 BGB.

[84]Article 6:101 PETL.

4.3 Robots and Roman Slaves

The slaves of ancient Rome are another category that could be instructional for our purposes.[85] While they were treated as "things" without rights or duties,[86] slaves were human beings with the same intellectual capacity as their masters.[87] In contrast to the justification for slavery on grounds of race found e.g., during the American slave era, slaves, at least in later periods of ancient Rome under the influence of the Stoics,[88] were not necessarily regarded as inferior, except perhaps socially or financially.[89] In that light, slaves may be likened to robots with full AGI capabilities.

Despite not having legal capacity to enter into binding agreements, slaves were an integral part of society. Slaves of the elite were often highly educated and used to conduct commerce on behalf of their master.[90] Indeed, slaves worked as "estate managers, bankers, and merchants."[91] An inevitable consequence of slaves taking part in general society was that slaves would cause harm to non-slaves both inside and outside the course of their duties. Roman law developed a sophisticated system to deal with such issues.

In the area of delict, this was the concept of *noxal* liability. The general rule was that a delict by a slave gave rise to an action against the master/owner (*dominus*) of the slave.[92] The master could also surrender the wrongdoer instead of paying damages, which effectively also limited the master's liability to the value of the slave.[93]

Another interesting concept was that of the *peculium*. Even though slaves regularly conducted commerce, a slave was a thing and any transactions entered into were unenforceable/invalid without having the master's authority or consent.[94] The peculium explains the peculiarity of slaves being such an integral part of commercial society despite their general incapacity to enter into legally binding agreements and even hold property.

The peculium was, in essence, a fund that masters allowed slaves (and their children)[95] to hold and, within limits, to deal with as de facto owners.[96] A party who had

[85] See for an overview Pagallo (2010), Katz (2010).

[86] Katz (2010).

[87] Bradley (1998), p. 478 et seq.

[88] Garnsey (1996), pp. 1–19.

[89] Bradley (1988), p. 478 et seq., citing Watson A (1987) Roman Slave Law, Johns Hopkins University Press, Baltimore and London; and p. 482. To be contrasted with Bradley (1988), pp. 481 et seqq, who argues that slaves in Ancient Rome were often regarded as inferior, similarly to how black slaves were in later times.

[90] Buckland (1908), p. 131; Katz (2010).

[91] Pagallo (2010), p. 400.

[92] See Buckland (1908), Chap. 5, pp. 98 et seqq. for an overview.

[93] Johnston (1995), p. 1525.

[94] Buckland (1908), p. 159 et seq.

[95] Johnston (1995), p. 1521.

[96] Buckland (1908), p. 187.

contracted with a slave could generally[97] enforce a judgment against the peculium, thereby adding a level of financial security to the transaction.[98] At the same time, a master's liability was limited to the amount of the peculium—an autonomous, albeit biological, limited liability company.[99]

Applying the principles of noxal liability and the peculium to robots, we get the basic rule that an owner would be liable for any harm caused by the robot; where this might be deemed excessive, one could introduce a financial limitation via a fund allocated to the robot.

4.4 Robots and Animals

Where slaves were sophisticated autonomous beings that the law treated as things but still granted a certain level of legal capacity, animals are autonomous beings lacking legal personhood, which is why the treatment of animals can serve as a rich source of parallels for how the law might or should treat robots,[100] especially in the early stages before robots achieve a general-purpose level of sophistication. Like with robots, humans' relationship to and use of animals is extremely diverse, as are the different species of animals. Animals can be trained or untrained; they can be domestic or wild; they can be kept for pleasure or they can be beasts of burden. These differences are important, as the law recognizes for the purposes of liability a meaningful distinction between the various human-animal relationships and types of animals.

For example, German tort law differentiates between luxury animals, such as pets, and domesticated animals that contribute to the economic livelihood of the owner, such as a sheep dog does for a shepherd. The owner is strictly liable for any harm caused by luxury animals. For the inherent, animal-specific harm caused by "domesticated economic animals," however, liability is like that of a supervisor for minors—a claim for liability may be defeated if the owner shows that they used the appropriate standard of care or that the harm would have occurred had such a standard been met.[101]

The law permits an exculpation for domesticated economic animals because it recognizes and balances their usefulness to society and their occupational necessity for the owner (which can be seen as a consequence of the fundamental right to occupational freedom guaranteed by Germany's Basic Law)[102] with the societal

---

[97]Not every act of a slave could give rise to an *actio de peculio*, however. See Johnston (1995), pp. 1522 et seqq.

[98]Johnston (1995), p. 1515 et seq.

[99]Katz (2010), Pagallo (2010), p. 400 et seq.

[100]Asaro (2011), p. 176; Schaerer et al. (2009), p. 73.

[101]§ 833 BGB.

[102]Article 12 GG. The right to occupational freedom could be restricted were the owner be subject to strict liability.

acknowledgement that accidents concerning such animals are to a certain extent unavoidable. What society demands is merely that due care was taken to avoid the accident. Luxury animals, however, serve no purpose beyond the pleasure they give their owner. Consequently, society has decided to subject the owner to strict liability for any resultant harm.

These concepts are not unique to German law. The common law recognizes two classes of animals: *ferae naturae* and *mansuetae naturae*.[103] The former are animals that are typically wild by nature, such as wolves; their keeper is generally subject to strict liability because of their inherent dangerous qualities.[104] The latter are animals that are typically domesticated and tame; their keeper will (generally)[105] only be held liable where the trait that led to the damage was or should have been known, the standard of care was not met, or where the keeper's negligence led to the damage. In determining whether the standard of care was met, the utility of keeping the animal, both to the keeper and to society at large, will be taken into consideration.[106]

Let us translate the above concepts to a shepherd and his dog. Our hypothetical plays on a calm mountain pasture in the Bavarian Alps. The dog is well-trained and has been herding sheep together with its owner for the past six years without incident. One day, however, while sitting next to the owner's nephew, the dog is stung by an errant wasp and bites the nephew, injuring him.

Applying the principles above, the shepherd would be able to defend against a claim for damages. He exercised the appropriate standard of care and the dog is used to contribute to his economic livelihood. Furthermore, the harm that occurred was a result of the animal-specific danger inherent to dogs—the possibility that a dog will bite.

We are easily able to replace the biological sheep dog of our hypothetical with a robot example. The question then is whether the law should decide differently.

4.5 Robots and Biology

Minors, agents, slaves, and animals all unite Calo's exceptional trio. They can physically alter their environment based on information (embodiment); their actions after assessing information cannot be predicted (autonomy); and they elicit emotional responses from humans (social valence). Thinking about other biological organisms, the same could be said for replicants or bioengineered intelligent animals, two possibilities that would be even closer to the original biological robots in Karel Čapek's play "R.U.R."

---

[103]McNeely (1939), p. 1182 et seq.

[104]McNeely (1939), p. 1183 et seq, p. 1208.

[105]Of course, exceptions exist, e.g., in the Cayman Islands a dog owner is strictly liable for any harm done to any person, cattle, or sheep by the dog. See Animals Law (2013 Revision), Section 39.

[106]McNeely (1939), p. 1198 et seq.

Considering, as we have seen, that the law successfully deals with mechanical robots' biological cousins, the question arises why the European Parliament and others are so quick to call for a special legal regime for robots. The Parliament gives as a reason the difficulty of being able to trace the cause of harm back to a specific human actor and of understanding whether the harmful act could have been avoided.[107] However, accountability where "many hands" are involved is not a problem unique to robots[108]; it is something the law regularly contends with.[109]

At first glance, the main difference between robots and their biological cousins is that the former are manufactured, while the latter organically come into existence with all the inexactitude nature exhibits (and which is an essential component of evolution). Animals are bred for desired characteristics and even humans have and could engage in eugenics, but manufacturing implies a level of precision and control over the outcome that is not yet feasible for biological production methods. However, upon coming into existence, both AIs and biological entities need to be conditioned, educated, trained, molded, etc.

A biological entity will typically have certain innate tendencies or instincts that need to be managed. Anybody who has spent time with animals can attest that individuals of the same breed and even litter can have vastly different personalities. Some dogs will be shy, others aggressive, and some friendly. The species of animal also makes a big difference; some are naturally curious and others are reclusive. As with humans, these tendencies can be encouraged or discouraged. A mistreated dog will be more likely to bite a human than one that has been raised with love.

Similarly, two robots from the same production line with the same base AI may end up taking vastly different actions depending on their training.[110] One line of action may be appropriate while another may lead to liability. For one, the dataset used to train the AI will influence how the robot ultimately acts. The UK's House of Lords Select Committee on Artificial Intelligence noted, "if the data is unrepresentative, or the patterns reflect historical patterns of prejudice, then the decisions which [robots] make may be unrepresentative or discriminatory as well."[111] For another, what a specific algorithm does with given data may also lead to bias,[112] as exemplified by Google's visual-recognition algorithms labeling black people in photos as gorillas.[113]

---

[107]See Footnote 60.

[108]See, for further examples and a discussion regarding computers and "many hands", Nissenbaum (1996), p. 29.

[109]Especially in regards to civil liability, which is the main problem discussed here. For a similar assessment, see Hubbard (2014), p. 1803 et seq., who also holds that existing liability schemes can provide for adequate solutions and provides some interesting argumentation in questioning the methodological background of approaches that demand a special treatment of the civil liability issues related to the activities of robots.

[110]Indeed, one proposed solution to address liability issues is to have certified "schools" for AIs. See European Parliament (2018), para. 56; GoodAI: https://www.goodai.com/school-for-ai. Accessed 25 May 2018.

[111]Lords Committee (2018), p. 41, para. 107.

[112]Lords Committee (2018), p. 42, para. 112.

[113]Simonite (2018).

Eventually, the algorithm used may be seen as similar to the selection of an animal with a specific personality or innate traits, while the dataset used may be more likened to the education and training imparted.

Regardless of the training required, how liability in certain situations is distributed is ultimately a societal choice that depends on the desired outcome, i.e., which actors should be privileged, etc. The distinction drawn by the law between luxury animals (pets) and privileged domesticated economic animals, as discussed above, highlights this very well.

The investigation so far has focused on drawing parallels between robots and already existing legal actors. But, if we acknowledge that the legal problems raised are neither particularly new, nor that they are particularly tied to the type of actor, we should take into account societal considerations for possible solutions. We should not follow our forebears' rules simply because the existing law can be adapted to a new development or because something has always been done so. Oliver Wendell Holmes Jr. called the latter "revolting," and believed that "if the training of lawyers led them habitually to consider more definitely and explicitly the social advantage on which the rule they lay down must be justified, they sometimes would hesitate where now they are confident, and see that really they were taking sides upon debatable and often burning questions."[114] As such, while "[f]or the rational study of the law the black-letter man may be the man of the present…the man of the future is the man of statistics and the master of economics."[115] Though neither statisticians nor economists, perhaps we can help raise awareness for a more considered approach to regulation.

5 To Whom Liability Shall Fall

Public policy touches upon a vastly wider realm of issues than just law. Regulation can guide societal developments, enabling and encouraging some while disabling others. This broader approach also applies to liability policy. From a societal perspective, liability generally has two goals: a corrective function of remedying harms and an incentive function of preventing harm.[116]

The corrective function can fail by misdirecting liability, i.e., attributing it to the wrong actor. Harm may be over-corrected insofar that strict liability will be pushed onto the manufacturer of the robot, regardless of whether the harm is one that could be expected and whether the manufacturer met their duty of care.[117] It may also be under-corrected in that a harm may end up uncompensated, for example because it is impossible to determine specifically what led to the harm or under who's charge

[114] Holmes (1897), p. 468.
[115] Holmes (1897), p. 469.
[116] Article 10:101 PETL. See also Nissenbaum (1996), p. 26 et seq.; Petit (2017), p. 20.
[117] Petit (2017), p. 21.

the robot was acting.[118] On the incentive side, an excessive emphasis on preventing harm beyond the reasonable could chill innovation.[119]

Our investigation above of general tort principles revealed some actors to whom liability could fall: to the robot itself, to the manufacturer, or to the owner/supervisor. In addition, it could also fall to no one; the damage would be fully sustained by the harmed. Which of these possibilities (and there are many more)[120] is the most suited must be assessed against best achieving liability's two goals.[121]

Let us start with the last-mentioned possibility of not holding anyone liable. This approach would fail both to correct any harms or incentivize the reduction of harms. As such, it should be rejected.

Liability falling to the robot itself would require that it is granted legal personhood.[122] There is no hindrance to such a step.[123] However, in order to be able to correct any harm, a robot would need a source of capital in order to pay damages—here, a personal fund tied to the robot, similar to a slave's peculium, could be one solution. Depending on its purpose, it could also serve the harmed party instead of paying monetary damages. Holding the robot itself liable could also meet the incentive function. For this, the robot would need to learn from the situations where it is held liable and adjust its behavior accordingly.[124]

Liability may also be accorded to the owner/supervisor of the robot. This approach would mirror more closely that of how the law currently responds to harms by autonomous entities, such as animals and agents. However, a user will rarely have the insight into the robot's working as would the manufacturer, which is an argument against this approach. Moreover, while the goal of remedying harms would be met, the incentive function would only be met where the market leads to such an outcome,

---

[118] Petit (2017), p. 20 et seq.

[119] Petit (2017), p. 17.

[120] Any number of actors in the "production chain" could be held liable. For example, Balkin (2015), p. 52, mentions "the owner, operator, retailer, hardware designer, operating system designer, or programmer(s), to name only a few possibilities."

[121] See also for a similar analysis of the below Eidenmüller (2017), p. 7 et seq.

[122] The legal literature explored this possibility already during the 1990 s. For such an early discussion, see Solum (1992), p. 1231 et seq., who goes as far as to consider a reimagining of the concept of legal personhood in order to accommodate the particularities of Artificial Intelligence agents.

[123] Hubbard (2011), p. 405 et seq., who puts forward some interesting ideas and tries to develop a test on how to assess in how far a robot might be able to be granted legal personhood.

[124] Robots are probably expected to learn in a practical way, either by adequate learning mechanisms integrated though their software or with some other type of intervention. On the contrary, one can raise doubts about the ability of robots to learn not to repeat the contested behavior through the classic mechanisms of enforcement of civil liability judgements. Even if robots are assigned property or even learn to obtain it themselves, they might not relate to it the way human agents do, as it might not have the same existential meaning for them. They might also not value their personal integrity the way human agents do. In that sense, the classic moral and material pressure that the traditional enforcement mechanisms, prescribed by civil procedural law, apply to human agents might not be completely emulated in the context of robotic behavior and thus the need for alternatives. For the moral and material stakes involved in the classic enforcement mechanisms of civil judgements, see among others Hazelhorst (2017), pp. 36–51. For the challenges of classic governmental enforcement mechanisms in the case of robots compare Morse (2018).

either through competition among manufacturers or because robot owners would have some sort of recourse against the manufacturers.

Picking up holding the manufacturer liable, we need to differentiate between the manufacturer of the robot and, where distinct, potentially the manufacturer of the AI as a component of the robot.[125] Recognizing the AI as a component is key for this approach. It is the robot manufacturer that combines and integrates the disparate parts into something that is more than their sum. It is also the robot manufacturer that is best situated to fully understand the complexities of the robot. The AI manufacturer will not necessarily control how the AI will be integrated into the robot.[126] Should the robot manufacturer be able to determine that it was the AI component that was the cause for any harm, it could still seek contractual regress.

In the EU, current product liability laws would make the manufacturers of the robot and of the AI jointly and severally liable.

Despite the complexities of the robot system and the difficulties of other parties in determining the cause for harm, it seems that existing liability patterns (e.g., strict liability or product liability) might be able to achieve the goals of harm correction and reduction that traditionally define the teleological horizon of the civil liability regimes.[127]

6 Conclusion

The law in general but also the law of extra-contractual civil liability in particular has traditionally been subjected to the question of how to regulate situations where liability could not, for all kinds of social, biological, or economic reasons, be located directly on the acting party. From the legal transactions of minors to the most recent development of the notion of product liability, regulators have proven an interesting adaptability, creativity, and resilience in dealing with the problem of assigning civil liability to appropriate actors so that the corrective and incentive functions that are traditionally associated with it survive within the context of new challenges. By the same token, the question of how to administer and distribute the risks of contemporary developments in order to secure a minimum level of justice has been shaped

[125]For the necessity of such delicate distinctions, especially between the different actors involved in the creation of the various hardware and software parts that constitute the robots see the interesting thoughts of Gurney (2013), pp. 247 et seqq., esp. pp. 258–266, which are made in the concomitant field of autonomous vehicles.

[126]That liability might predominantly lie with the manufacturer see also Marchand and Lindor (2012), pp. 1326–1330.

[127]While strict liability does carry the risk of over-correcting harm, any such over-correction can be addressed through supplemental measures, such as a liability fund for manufacturers of robots and AI components, mandatory insurance for the users of robots, testing and certification schemes that could limit liability, etc. Such traditional regulatory schemes can, in any case, serve as a platform on which to develop an adequate solution. It is the duty of legal literature and, of course, that of case law, to progressively define the necessary details.

by previous leaps in technological advancements.[128] As a result, legal orders can claim to be in a position to recognize potential conflicts and identify basic regulatory patterns that provide for some form of methodological guidance whenever technology shakes the established socio-economic structures that define a particular period, paving the way for societal disruption and transformation. There seems to be no compelling reason to deviate from such known mechanisms in dealing with the extra-contractual liability of robots. Nor does it seem necessary to completely overthrow the established methodological arsenal developed in previous encounters with technologically driven changes to the established social and economic paradigm. Nonetheless, one shall be very careful when making such a statement, as the outer limits of any methodological approach need to be clearly defined in order to avoid misinterpretations. By placing trust in certain traditional extra-contractual civil liability mechanisms, this chapter does not imply that pouring new wine into old bottles is the proper methodology to secure an adequate solution to the problem of the liability of robots. But, if legal methodology has learned something from dealing with the Internet as a similarly disruptive technological advancement like the one that now arises from the embodied applications of artificial intelligence, it would be that the unique characteristics of new technological developments might not necessarily harm or even rule out regulatory efforts.[129] On the contrary, they might be part of the solution and through their interplay with traditional regulatory mechanisms and methodologies form a new regulatory reality, [130] where the distinctive elements of the new are neatly incorporated into the structures of the old, so that both are respected for the sake of justice. By adequately incorporating our new intelligent creations in our traditional regulatory framework we might, after all, inspire them to remain our partners and not opt to become our dystopian overlords.[131]

References

Animals Law. (2013 Revision). Cayman Islands Supplement No. 1 published with Extraordinary Gazette No. 82 of 11 October 2013.

Asaro, P. (2011). A body to kick, but still no soul to damn: Legal perspectives on robotics. In P. Lin, K. Abney, & G. Bekey (Eds.), *Robot ethics: The societal and social implications of robots*. Cambridge: MIT Press.

Balkin, J. M. (2015). The path of robotics law. *California Law Review Circuit, 6,* 45–60.

Beck, U. (1986). *Risikogesellschaft. Auf dem Weg in eine andere Moderne*. Frankfurt: Suhrkamp.

[128] Beck (1986).

[129] See just indicatively and in the context of internet regulation, Mayer (1996), p. 1782 et seq.; Reidenberg (1998), p. 553 et seq.; Goldsmith (1999), p. 1199 et seq.; Paulus (1999), p. 443 et seq.; Lessig (1999), p. 501 et seq.; Nachbar (2000), p. 214 et seq.; Perritt (2001), p. 215 et seq.; Hughes (2003), p. 359 et seq.; Geist (2003), p. 323 et seq.

[130] Compare again within the context of the regulation of the Internet, Reed (2012), pp. 49–83, 105–128, 151–178.

[131] For the moral aspects of the relationship between humans and robots, see Coeckelbergh (2011).

Bostrom, N. (1998). How long before superintelligence? *International Journal of Future Studies*, 2. Available with updates https://nickbostrom.com/superintelligence.html. Accessed May 25, 2018.
Bradley, K. (1988). Roman slavery and roman law. *Historical Reflections/Réflexions Historiques, 15*(3), 477–495.
Brooks, R. (1991). Intelligence without reason. Massachusetts Institute of Technology, Artificial Intelligence Laboratory, A.I. Memo No. 1293, April 1991, prepared for Computers and Thought, ICJAI-91, http://hdl.handle.net/1721.1/6569. Accessed May 25, 2018.
Buckland, W. (1908). *The roman law of slavery: The condition of the slave in private law from augustus to justinian*. Cambridge: Cambridge University Press.
Calo, R. (2015). Robotics and the lessons of cyberlaw. Legal studies research paper no. 2014–08. *California Law Review, 103,* 513–563.
Coeckelbergh, M. (2011). Humans, animals, and robots: A phenomenological approach to human-robot relations. https://link.springer.com/article/10.1007/s12369-010-0075-6. Accessed May 25, 2018.
Easterbrook, F. H. (1996). Cyberspace and the law of the horse. *University of Chicago Legal Forum, 1996*(7), 207–216.
Eidenmüller, H. (2017). The rise of robots and the laws of humans. Oxford Legal Studies Research Paper No. 27/2017. https://papers.ssrn.com/sol3/papers.cfm?abstract_id=2941001. Accessed May 25, 2018.
European Group on Tort Law (EGTL) Principles of European Tort Law. http://civil.udg.edu/php//index.php?id=283. Accessed May 25, 2018.
European Group on Tort Law (EGTL) Principles of European Tort Law—A Harmonization Project. http://civil.udg.edu/php//index.php?id=300. Accessed May 25, 2018.
Fitzpatrick, K. K., Darcy, A., & Vierhile, M. (2017). Delivering cognitive behavior therapy to young adults with symptoms of depression and anxiety using a fully automated conversational agent (Woebot): A randomized controlled trial. *JMIR Mental Health, 4*(2), e19.
Garnsey. (1996). *Ideas of slavery from aristotle to augustine*. Cambridge: Cambridge University Press.
Geist, M. (2003). Cyberlaw 2.0. *Boston College Law Review, 44,* 323–358.
Gibbs, S. (2015, July 27). Musk, Wozniak and Hawking urge ban on warfare AI and autonomous weapons. *The Guardian*. https://www.theguardian.com/technology/2015/jul/27/musk-wozniak-hawking-ban-ai-autonomous-weapons. Accessed May 25, 2018.
Goldsmith, J. (1999). Against cyberanarchy. Occasional Papers from the Law School of the University of Chicago, No. 40.
Goldstein, J. (1999). Emergence as a construct: History and issues. *Emergence, 1*(1), 49–72.
Gurney, J. K. (2013). Sue my car not me: Products liability and accidents involving autonomous vehicles. *University of Illinois Journal of Law, Technology and Policy, 2013*(2), 247–277.
Hartnett, K. (2018a). Smart swarms seek new ways to cooperate. *Quanta Magazine*. https://www.quantamagazine.org/smart-swarms-of-robots-seek-better-algorithms-20180214/. Accessed May 25, 2018.
Hartnett, K. (2018b). The simple algorithm that ants use to build bridges. *Quanta Magazine*. https://www.quantamagazine.org/the-simple-algorithm-that-ants-use-to-build-bridges-20180226/. Accessed May 25, 2018.
Hazelhorst, M. (2017). *Free movement of civil judgments in the European Union and the right to a fair trial*. T.M.C. Asser Press, The Hague.
Holmes, O. W. (1897). The path of the law. *Harvard Law Review, 10*(8), 457–478.
Hubbard, F. P. (2011). "Do androids dream?": Personhood and intelligent artifacts. *Temple Law Review, 83,* 405–474.
Hubbard, F. P. (2014). Sophisticated robots: Balancing liability, regulation, and innovation. *Florida Law Review, 66*(5), 1803–1872.
Hughes, J. (2003). The internet and the persistence of law. *Boston College Law Review, 44,* 359–396.
Johnson, S. (2001). *Emergence—The connected lives of ants, brains, cities, and software* (eBook ed.). New York: Scribner.

Johnston, D. (1995). Limiting liability: Roman law and the civil law tradition. *Chicago-Kent Law Review, 70,* 1515–1538.
Judson, O. (2017). What the octopus knows, The Atlantic, January/February 2017 Issue. https://www.theatlantic.com/magazine/archive/2017/01/what-the-octopus-knows/508745/. Accessed May 25, 2018.
Katz, A. (2010). Intelligent agents and internet commerce in ancient Rome, society for computers and law. https://www.scl.org/articles/1095-intelligent-agents-and-internet-commerce-in-ancient-rome. Accessed May 25, 2018.
Koch, B. (2007). The "principles of European Tort Law". *ERA-FORUM 2000, 8*(1), 107–124.
Lessig, L. (1999). The law of the horse: What cyberlaw might teach. *Harvard Law Review, 113,* 501–546.
Lords Committee. (2018). AI in the UK: Ready, willing and able? House of Lords Select Committee on Artificial Intelligence, Parliament of the United Kingdom. Report of Session 2017–19, published April 16, 2017, HL Paper 100. https://publications.parliament.uk/pa/ld201719/ldselect/ldai/100/100.pdf. Accessed May 25, 2018.
Marchand, G. E., & Lindor, R. A. (2012). The coming collision between autonomous vehicles and the liability system. *Santa Clara Law Review, 52,* 1321–1340.
Markoff, J. (2011). Computer wins on 'Jeopardy!': Trivial, It's not. *The New York Times.* https://www.nytimes.com/2011/02/17/science/17jeopardy-watson.html. Accessed May 25, 2018.
Mayer. (1996). Recht und cyberspace. *Neue Juristische Wochenschrift, 48*(28), 1782–1791.
McNeely, M. (1939). A footnote on dangerous animals. *Michigan Law Review, 37*(8), 1181–1208.
Morse, S. (2018). Government-to-robot enforcement. https://papers.ssrn.com/sol3/papers.cfm?abstract_id=3143716. Accessed May 25, 2018.
Nachbar, T. (2000). Paradox and structure: Relying on government regulation to preserve the internet's unregulated character. *Minnesota Law Review, 85,* 215–318.
Nissenbaum, H. (1996). Accountability in a computerized society. *Science and Engineering Ethics, 2*(1), 25–42.
Pagallo, U. (2010). The human master with a modern slave? Some remarks on robotics, ethics, and the law. In M. Arias-Oliva, et al. (Eds.), *Proceedings of the Eleventh International Conference.* The "backwards, forwards and sideways" changes of ICT, ETHICOMP 2010, 14–16 April 2010. Universitat Rovira i Virgili: Tarragona.
Paulus, C. (1999). Multimedia: Herausforderung an das Wirtschaftsrecht. *Multimedia und Recht, 2*(8), 443–447.
Perrit, Jr. H. (2001). Towards a hybrid regulatory scheme for the internet. The University of Chicago Legal Forum 1: Article 8.
Petit, N. (2017). Working paper, Law and regulation of artificial intelligence and robots: Conceptual framework and normative implications (March 9, 2017). https://papers.ssrn.com/sol3/papers.cfm?abstract_id=2931339. Accessed May 25, 2018.
Privacy International, ARTICLE 19. (2018). Report: Privacy and freedom of expression in the age of artificial intelligence, April 2018. https://privacyinternational.org/report/1752/privacy-and-freedom-expression-age-artificial-intelligence. Accessed May 25, 2018.
Reed, C. (2012). *Making laws for cyberspace.* Oxford: Oxford University Press.
Reidenberg, J. (1998). Lex informatica: The formulation of information policy rules through technology. *Texas Law Review, 76*(3), 553–593.
Saigle, V., Dubljević, V., & Racine, E. (2018). The impact of a landmark neuroscience study on free will: A qualitative analysis of articles using libet and colleagues' methods. *AJOB Neuroscience, 9*(1), 29–41.
Schaerer, E., Kelley, R., & Nicolescu, M. (2009). Robots as animals: A framework for liability and responsibility in human-robot interactions. The 18th IEEE international symposium on robot and human interactive communication, September 27–October 2, 2009, IEEE, Toyama.
Scherer, M. (2016). Regulating artificial intelligence systems: Risks, challenges, competencies, and strategies. *Harvard Journal of Law & Technology, 29*(2), 354–400.

Simonite, T. (2018). When it comes to gorillas, Google Photos remains blind, Wired, January 11, 2018. https://www.wired.com/story/when-it-comes-to-gorillas-google-photos-remains-blind/. Accessed May 25, 2018.

Solum, L. B. (1992). Legal personhood for artificial intelligences. *North Carolina Law Review, 70*, 1231–1287.

Urban, T. (2015). The AI revolution: The road to superintelligence, wait but why. https://waitbutwhy.com/2015/01/artificial-intelligence-revolution-1.html. Accessed May 25, 2018.

Vladeck, D. (2014). Machines without principals. *Washington Law Review, 89*(1), 117–150.

Wolchover, N. (2017). 'Digital alchemist' seeks rules of emergence. *Quanta Magazine*. https://www.quantamagazine.org/digital-alchemist-sharon-glotzer-seeks-rules-of-emergence-20170308/. Accessed May 25, 2018.

Yong, E. (2010). Brainless slime mould makes decisions like humans. *Discover Magazine*. Not Exactly Rocket Science Blog. http://blogs.discovermagazine.com/notrocketscience/2010/08/10/brainless-slime-mould-makes-decisions-like-humans/. Accessed May 25, 2018.

Zittrain, J. (2006). The generative internet. *Harvard Law Review, 119*, 1974–2040.

Business and Regulatory Responses to Artificial Intelligence: Dynamic Regulation, Innovation Ecosystems and the Strategic Management of Disruptive Technology

Mark Fenwick, Erik P. M. Vermeulen and Marcelo Corrales

Abstract Identifying and then implementing an effective response to disruptive new AI technologies is enormously challenging for any business looking to integrate AI into their operations, as well as regulators looking to leverage AI-related innovation as a mechanism for achieving regional economic growth. These business and regulatory challenges are particularly significant given the broad reach of AI, as well as the multiple uncertainties surrounding such technologies and their future development and effects. This chapter identifies two promising strategies for meeting the "AI challenge," focusing on the example of Fintech. First, "dynamic regulation," in the form of regulatory sandboxes and other regulatory approaches that aim to provide a space for responsible AI-related innovation. An empirical study provides preliminary evidence to suggest that jurisdictions that adopt a more "proactive" approach to Fintech regulation can attract greater investment. The second strategy relates to so-called "innovation ecosystems." It is argued that such ecosystems are most effective when they afford opportunities for creative partnerships between well-established corporations and AI-focused startups and that this aspect of a successful innovation ecosystem is often overlooked in the existing discussion. The chapter suggests that these two strategies are interconnected, in that greater investment is an important element in both fostering and signaling a well-functioning innovation ecosystem and that a well-functioning ecosystem will, in turn, attract more funding. The result-

M. Fenwick (✉)
Graduate School of Law, Kyushu University, Fukuoka, Japan
e-mail: mdf0911@gmail.com

E. P. M. Vermeulen
Department of Business Law, Tilburg University, Tilburg, The Netherlands

E. P. M. Vermeulen
Legal Department, Philips Lighting, Eindhoven, The Netherlands

M. Corrales
Center for Innovation and Research, Universidad Politécnica y Artística del Paraguay (UPAP), Asunción, Paraguay

M. Corrales et al. (eds.), *Robotics, AI and the Future of Law*, Perspectives in Law, Business and Innovation, https://doi.org/10.1007/978-981-13-2874-9_4

ing synergies between these strategies can, therefore, provide a jurisdiction with a competitive edge in becoming a regional hub for AI-related activity.

Keywords AI ecosystems · Dynamic regulation · Experimentation
Innovation ecosystems · Regulatory sandboxes · Venture capital

1 Introduction

Finding an appropriate response to disruptive new AI technologies is enormously challenging for any business looking to integrate AI into their operations or for regulators looking to leverage AI-related innovation as a means of achieving regional economic growth. These business and regulatory challenges are particularly significant given the potential reach of AI, as well as the multiple uncertainties surrounding such technologies and their future development and effects.

This chapter begins with a brief overview of three key features of the "AI challenge" (Sect. 2). The three features identified are (i) the disruption of traditional business models triggered by AI-technologies (Sect. 2.1); (ii) the increase in AI-driven investment and the new opportunities and resulting disruption that this has triggered (Sect. 2.2); and (iii) the profound uncertainties that surround the possible future development and effects of AI-related technologies (Sect. 2.3). Each of these issues has significant implications for business and regulators.

Meeting the AI challenge is crucial for established corporations, startups and policymakers. The chapter identifies two promising strategies for regulating AI based on the experience of regulating previous disruptive new technologies. First, "dynamic regulation," in the form of so-called "regulatory sandboxes" and other proactive regulatory approaches, that aim to provide a space for responsible AI-related business innovation. An empirical study provides preliminary evidence to suggest that jurisdictions that adopt a more proactive approach to the regulation of Fintech do indeed seem to attract greater investment. Section 3 outlines some of the main features of such an approach.

The second regulatory strategy relates to so-called "innovation ecosystems." It is suggested that such ecosystems are particularly effective when they afford opportunities for more creative partnerships between established corporations and AI-focused startups and that this feature of ecosystems is often neglected in the existing discussion. The main features and potential benefits of such partnerships are outlined in Sect. 4.

The chapter argues that these two strategies are inter-connected, in that greater investment is an important element in fostering a well-functioning innovation ecosystem. The resulting synergies between these two strategies can, therefore, provide a jurisdiction with a competitive edge in an effort to become a regional hub for AI-related activities.

Although these regulatory strategies are not riskless (either for business or the state), they do, nevertheless, represent the best option for responding to the AI challenge. At least, they seem to be clearly preferable to the two obvious alternatives, namely strict ex ante control (which risks stifling innovation, investment, and growth, or—at least—prompting a "brain drain" or capital flight) or a socially-irresponsible de-regulation (which may result in harmful effects, particularly in the context of technologies whose effects are uncertain).

2 The "AI Challenge"

It is helpful to begin by distinguishing three important aspects of the "AI challenge," at least as it impacts upon business and government (regulators and other policy-makers).

2.1 AI-Technologies & the Disruption of Existing Business Models

When thinking about the business and regulatory challenges that are created by AI technologies, it is important to adopt a broad-based definition of AI. Such a definition encompasses all four main types of AI technology:

i. "Type 1" AI refers to purely "reactive" machines that specialize in one area or task. For instance, the drafting and review of loan agreements. More "famous" examples would be IBM's Deep Blue chess software or Google's Alpha Go algorithm for playing Go;
ii. "Type 2 AI" machines possess just enough memory or "experience" to make proper decisions and execute appropriate actions in specific situations or contexts. Self-driving cars, chatbots, or personal digital assistants are the most commonly cited examples;
iii. "Type 3" AI has the capacity to understand thoughts and emotions which affect human behavior. Softbank Robotic's "Pepper" can organize large amounts of data and information in order to have a "human-like" conversation;
iv. "Type 4 AI" is "artificial intelligence" as it is typically portrayed in Hollywood movies or TV shows (think HBO's Westworld). Machines using this type of AI are self-aware, super-intelligent, sentient and are presumed to possess something like consciousness.

The advantage of an expansive definition of this kind is that it highlights the urgency of AI-related developments and avoids the risks of complacency. If we define AI narrowly in terms of Type 4 AI—i. e., AI that is "more human than human"—we don't need to be overly concerned with the disruptive impact and potential effects of such technologies at the moment. After all, "singularity"—the moment when AI capacities surpass our own—seems to be some decades away. The advantage of a broader definition of AI is that it, therefore, allows us to appreciate the extent and diversity of the business challenges that are *already* created by AI and allows us to develop a portfolio of possible regulatory strategies appropriate for managing such technologies. Moreover, all aspects of business operations seem likely to be affected by at least one of these four types of AI, and such disruption is already occurring across multiple sectors of the economy.

Take the example of financial services. There are many tasks that are central to financial services that can already be better performed by machines. Type 1 AI can do certain things more effectively than a human (for instance, reviewing large numbers of standard form contracts). Even though Type 4 AI may be a long way off that doesn't mean that the financial services industry isn't already being disrupted by other, "simpler," forms of machine intelligence.

Fintech—broadly defined as the use of new technologies to make financial services, ranging from online lending to digital currencies, more efficient—can already be seen across a range of financial services and many of these new services involve some kind of AI, in the broad sense adopted here.[1] For example, p*eer-to-peer lending platforms that use algorithms and machine learning to assess the creditworthiness of borrowers. Or, "robo-advisors" that automate many aspects of personal finance and wealth management. Such intelligent machines can already help individuals manage their personal accounts, debts, assets and investments.*[2]

Or, in the context of healthcare and life sciences: artificial intelligence already takes the role of an experienced clinical assistant that can help doctors make faster and more reliable diagnoses. We already see AI applications in the areas of imaging and diagnostics, and oncology, for example.[3] More generally, machine learning has the potential to improve remote patient monitoring. AI algorithms are able to take information from electronic health records, prescriptions, insurance records and even wearable sensor devices to design a personalized treatment plan for patients. Finally, AI-related technologies accelerate the discovery and creation of new medicines and drugs. There is a broad consensus amongst insiders that healthcare is being transformed for the better as a result of AI. And the opportunities and potential are limitless.

Broadly defined, it is hard to imagine any existing business that isn't profoundly disrupted by AI. Artificial intelligence, machine learning, and deep learning are just the beginning of a revolution that will transform everyday life and how we interact with technology. And new AI-oriented start-ups looking to satiate this new demand are rapidly emerging, which brings us to the second aspect of the AI challenge.

---

[1]For a general introduction to Fintech, see Arner et al. (2016a, b), Haddad and Hornuff (2016) and MIT Sloan School of Management (2016). For a discussion of Fintech and investment-related issues and trends, see Fenwick et al. (2018).

[2]See, generally, Fenwick and Vermeulen (2017).

[3]For a general introduction, see JASON (2017).

2.2 AI-Driven Investment, Start-Ups & a New Market for Corporate Control

The result of these technological and business developments is that AI is attracting record amounts of investors' money. Consider global venture capital investments in AI, which increased significantly between 2012 and 2016, both in terms of absolute amount invested and the number of deals (see Fig. 1).

Moreover, M&A activity involving AI companies has increased significantly over a similar five-year time-scale (see Fig. 2). Typically, the acquired companies in such acquisitions are often Silicon Valley-based startups and the majority are from the U.S. (see Fig. 3). But that doesn't necessarily make it the U.S. the only AI center in the world. Even a high-level review of the available data suggests that other regions are active and that AI entrepreneurs can be found everywhere.

It seems clear that established corporations and investors value new companies that embrace these new technologies and gradually bring them to market. It is hardly surprising that have replaced the financial institutions and oil businesses as the largest companies in the world, at least according to their market capitalization.

And if we consider some of the world's largest companies—think Apple, Alphabet, Microsoft or Amazon—we can see that these companies view different types of Artificial Intelligence as a key business opportunity for the future. Siri (Apple), Google Assistant (Alphabet), Cortana (Microsoft) and Alexa (Amazon) are already able to assist you with more and more difficult tasks, and this trend is only set to continue.

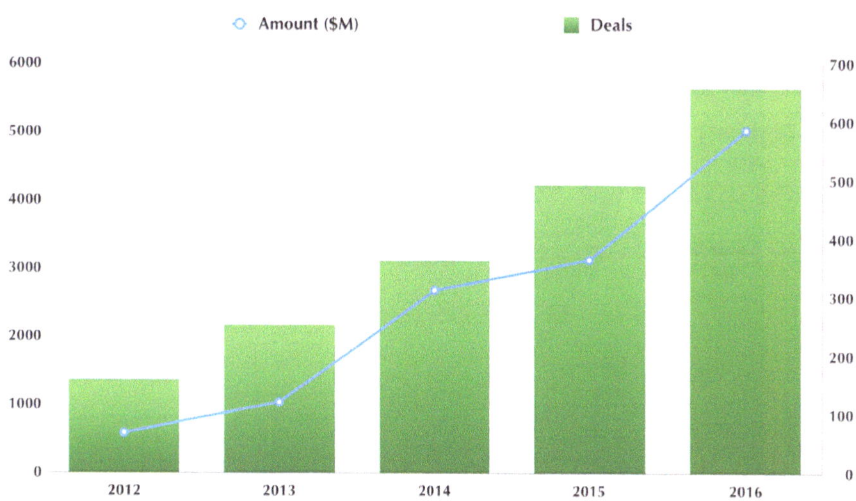

Fig. 1 Global venture capital investments in AI by amount & no. of deals (2012–16). *Source* CB insights

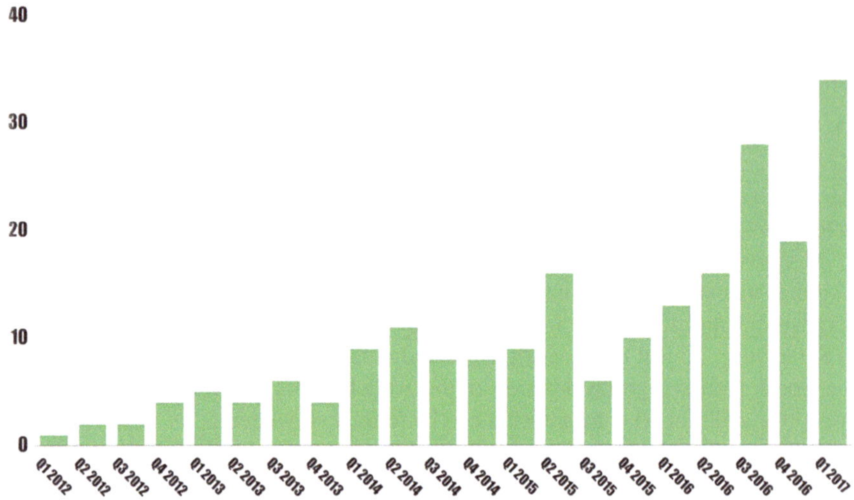

Fig. 2 M&A activity involving AI companies by number of deals (2012–17). *Source* CB insights

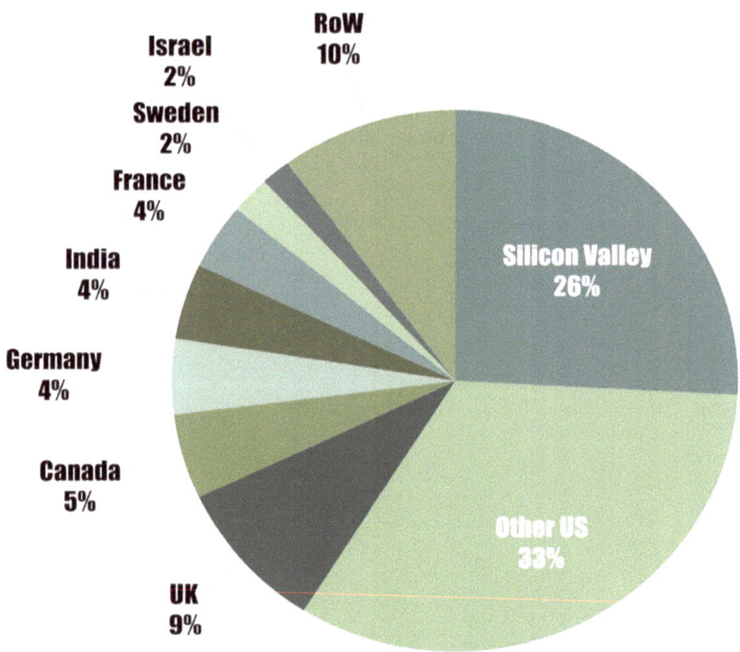

Fig. 3 The location of acquired AI companies (2012–2017). *Source* Pitchbook

2.3 AI Technologies & Radical Uncertainty

A final point regarding the challenge of AI technologies. As Types of AI develop more uncertainties will inevitably occur, particularly as we approach "singularity", and this highlights a transformation in the character of risk. A paradox of digital technologies is that they make our lives easier, but they also make the world harder—perhaps even impossible—to understand. The digital world is a world of risk—of identifiable and measurable dangers—but, more significantly, it is also a world of radical uncertainty.[4] Our relationship with new technology is often characterized by uncertainty, in the sense that "all we know is that there are many things that we do not know" about a technology and its effects.

The speed of technological development means that transformative change will come much sooner than expected. Big data and the near-endless amounts of information have undoubtedly transformed AI to unprecedented levels. Blockchain technology and smart contracts will merely continue and, very probably, accelerate the trend. The enormous increase in computational power, the breakthrough of "Internet of Things" applications and the further development of smart machines will only accelerate AI's development and global adoption. The acceleration of innovation will add to AI's ability to adapt to new situations and solve problems that currently seem to be impossible.

Any list of potential outcomes—positive or negative—created by new technologies is always going to be incomplete. As such, the digital world is a world where "reality" and "truth" regarding new technologies are uncertain, unsettled and constantly being contested.

In part, this is simply a function of the ever-quickening speed of technological change.[5] As soon as we believe that we have a clear understanding, a new development has already occurred that renders any existing understanding obsolete. But something else is also going on. "Understanding" of complex man-made systems is now increasingly "beyond" human comprehension.[6] For the first time in history, we live in a world where more and more technologies are simply beyond human comprehension.

So how can business and regulators meet this AI challenge? In this chapter, two potentially fruitful strategies are introduced and explored. First, new forms of regulation, notably regulatory sandboxes (Sect. 3), and, second, innovation ecosystems that foster partnerships between established firms and startups (Sect. 4).

[4]See Beck (1992).

[5]See Bennett Moses (2011).

[6]See Arbesman (2016).

3 "Responsive"/"Smart"/"Dynamic" Regulation

Recently, the regulation literature—particularly that branch of the discussion focusing on the regulation of new and disruptive technologies has focused on improving the ability of regulators to respond to changing industry practices (especially technology-driven changes) and the ability to improve relationships between regulators and regulated companies.[7] So-called dynamic regulation can respond to changing industry practices through feedback effects and enhanced information for regulation.

In this context, what is particularly important is that that within the framework of these new and more dynamic models, regulatory decisions should not be thought of as 'final events' (to be made for all-time and from which we "all move on"). Rather, we should think of regulatory choices as a form of "measured decision-making," i.e., regulatory choices are open-ended and highly contingent selections that are merely one stage or element in a longer narrative and not the "final word" on a particular issue. As such, regulators need to abandon a fixation on finality and legal certainty and embrace contingency, flexibility and an openness to the new. The justification for this new openness derives from the contingency of the technology-dominated environment in which regulators now must operate.

This shift in perspective affects how we regulate disruptive technologies. Rather than approaching decisions as "final events" (to be made for all-time and to which we all commit), Michel Callon has proposed the alternative notion of "measured action" (i.e., measured decision-making), where you do not decide an outcome, you take contingent measures that are based on inclusive processes involving both experts and the public.[8] Any regulatory "choice" ultimately remains open-ended, leaving space to incorporate new knowledge, discoveries, and claims. The need for finality, Callon argues, is usually overstated, more the product of expediency and habit than actual necessity.

Similarly, Gralf-Peter Caliess and Peer Zumbasen's concept of "rough consensus and running code" developed in the context of transnational business law also highlights a new contingency as a defining feature of contemporary "law"-making in transnational settings.[9]

According to this perspective, the antidote to the "Hobson's choice" of recklessness (i.e., an irresponsible deregulation or non-regulation) versus paralysis (i.e., an excess of regulation that stifles innovation) in the regulation of disruptive technologies is a willingness to remove the temporal horizon that has traditionally defined decision-making, while at the same time creating new and more dynamic mechanisms for consistent citizen involvement in the ongoing process of determining measured action.

Of course, this may very well be a noble goal, but the problem is how to operationalize such an approach in more concrete terms. An obvious solution to this regulatory dilemma might be to adopt some form of policy experimentation, i.e.,

[7]See, e.g., Black et al. (2007), Black (2009) and Kaal (2013, 2014).

[8]Callon et al. (2009).

[9]Calliess and Zumbansen (2010).

testing different regulatory schemes and then comparing the results. But such experimentation poses a problem for regulators. Too often, regulators define "success" in negative terms, as in the avoidance of catastrophe:

> One defining feature of the Strong Precautionary Principle is that it places a governmental entity in a role as a risk gatekeeper. Implicit in the Principle is the idea that there must be a 'decider' who will determine whether the proponent of the activity has met its burden of proof on safety. The preventive thrust of Strong Precaution further implies that this review of risks should occur *before* the activity commences or the potentially risky product reaches the market.[10]

Avoiding grounds for criticism inevitably results in an overly cautious approach, often called the "precautionary principle." In this regard, the recent "rise" of the so-called "regulatory sandbox" is particularly interesting as a concrete approach to regulation that ensures a responsible regulatory framework that doesn't have a chilling effect on technological innovation.

3.1 Regulatory Sandboxes

In the financial industry, it has recently been suggested that such an engaged-approach with disruptive technology is best facilitated by the establishment of regulatory sandboxes.

The Financial Conduct Authority (FCA), the financial regulatory body in the United Kingdom, is widely accredited with first introducing this approach.[11] In April 2016, the FCA broke new ground by announcing the introduction of a regulatory sandbox which allows both start-up and established companies to "test" new ideas, products and business models in the area of Fintech.

The aim of the sandbox is to create: a "safe space" in which businesses can test innovative products, services, business models and delivery mechanisms without immediately incurring all the normal regulatory consequences of engaging in the activity in question.

The idea behind the sandbox is for the state regulator to approve a firm-specific, de-regulated space for the testing of innovative products and services without being forced to comply with the applicable set of existing rules and regulations. With the sandbox, the regulator aims to foster innovation by lowering regulatory barriers and costs for testing disruptive innovative technologies, while ensuring that consumers will not be negatively affected. The three key questions that were investigated by the FCA on the sandbox proposal concerned "regulatory barriers" (how and to what extent can they be lowered?), "safeguards" (what protector measures should be in place to ensure safety), and "legal framework" (what regulatory arrangement is mandated by EU law).

---

[10] Sachs (2011), p. 1298; see also Sunstein (2005).

[11] See Financial Conduct Authority (2015).

What is perhaps most interesting about the sandbox is that new ideas, products and services can be tested in a "live," "real world" environment. As such, firms are given the authorization to test their products or strategies without being subjected to existing regulatory requirements and the associated prohibitions or compliance costs. In order to create this environment, the FCA has defined a set of default parameters that can be altered on a case-by-case basis, depending on the needs of a particular firm. These parameters include:

i. *Duration.* As a default, the FCA considers three to six months to be an appropriate length of time to 'test' a particular innovation.
ii. *Scale.* The number of customers should be big enough to generate statistically relevant data and information on the product or service. This means that customers should be selected based on certain criteria that are appropriate for the product and service. Clearly, pre-agreed safeguards and protections should be in place to protect consumers.
iii. *Prior Disclosure.* Customers should be accurately informed about the test and any available compensation (if needed). Moreover, indicators, parameters and milestones that are used during the testing phase should be clearly set out from the outset.

What makes the regulatory sandbox model so attractive is that, insofar as technology has consequences that flow into everyday lives, such technology will be open to discussion, democratic supervision and control. In this way, public entitlement to participate in regulatory debates can help to create a renewed sense of legitimacy and confidence that justifies the regulation that is subsequently adopted.

It comes as no surprise that regulatory sandboxes are being adopted by other regulators, such as the Australian Securities and Investment Commission, Singapore's Monetary Authority and Abu Dhabi's Financial Services Regulatory Authority.

In discussions about regulatory sandboxes with other experts in banking and finance, arguments are often made suggesting that their deployment is nothing more than a strategy of a country to signal its openness to innovation and technology. In their view, "sandboxes" aren't really offering anything new. Regulators are usually able to exempt companies and technologies to comply with the applicable set of rules and regulations without referring them to a sandbox. The Australian "Fintech" exemption is an example.

Yet, these arguments seem to miss the main advantages of the "regulatory sandbox." The potential of regulatory sandboxes goes much further than a signal. Insofar as technology has consequences that flow into everyday lives, such technology will be open to discussion and democratic supervision and control. In this way, public entitlement to participate in regulatory debates can help to create a renewed sense of legitimacy that justifies the regulation.

What is even more important is that regulatory sandboxes offer opportunities to generate information and data relevant to the regulation of the new digital world. They allow the participants in the sandbox, i.e., regulators, incumbent companies, start-ups, investors, consumers, to learn about the new technologies (such as AI). In this way, they can create the necessary dialogue that helps us understand new technologies.

They allow for collaboration and joint discovery. But perhaps most importantly, they create an opportunity to change the mind-set of incumbents operating in the financial service sector and allow them to embrace the new possibilities associated with artificial intelligence, machine learning and deep learning.

In an age of constant, complex and disruptive technological innovation, knowing prioritized *what*, *when*, and *how* to structure regulatory interventions has become much more difficult. Regulators can find themselves in a situation where they believe they must opt for either reckless action (regulation without sufficient facts) or paralysis (doing nothing). Inevitably in such a case, caution tends to be over risk. The precautionary principle becomes the default position. But such caution merely functions to reinforce the status quo and the result is that new technologies struggle to reach the market in a timely or efficient manner.

3.2 An Empirical Test

In order to explore empirically the effects of a more dynamic regulatory approach to disruptive new technologies, a small empirical study was conducted, focusing on the example of Fintech.

Broadly speaking, if we look around the world today we can distinguish between two broad categories of regulatory response—"reactive" and "proactive"—both of which comprise a number of sub-categories.

On the one hand, there are what we can characterize as reactive jurisdictions. This includes countries in which nothing is being done, i.e., there is currently no regulatory talk or action responding to Fintech. A second sub-group consists of those countries in which there is only partial or fragmented regulation of Fintech. Certain institutions, such as the Consumer Financial Protection Bureau in the United States, may offer certain safe harbor provisions for certain type of Fintech companies. Yet, there appears to be little willingness to genuinely embrace the technology and its regulatory implications, nor is there any comprehensive plan as to how Fintech can or should be regulated.

On the other hand, are those countries that take a more a proactive approach. In this group, we find those countries that make Fintech a strategic priority. In such countries, more regulatory attention paid to Fintech. Again, a number of sub-categories can be identified.

A first sub-group comprises countries in which such "attention" takes the form of consultation papers, White Papers, or conferences. Of course, there is a risk that such "talk" can slide into empty "lip service" aimed at projecting an image of regulatory action when, in reality, action is limited or non-existent.

A second sub-group of countries engage in what we might characterize as "regulatory guidance." Regulators issue guidelines or provide advice to Fintech start-ups and incumbents in order to help navigate them through the regulatory system. This does not necessarily entail changes in formal regulatory structures, but it does provide some support innovation. The initiative to issue a national charter for the supervision of Fintech companies by the U.S. Office of the Controller of the Currency is a recent example.

A final group of countries has embraced the possibilities of Fintech by creating a regulatory sandbox, as described above. We would characterize this approach as "regulatory experimentation." Regulators create a sandbox in which they facilitate and encourage a space to experiment. This allows the testing of new technology-driven services, under the supervision of regulators. This ensures that meaningful data can be gathered for the evaluation of risk in a safe environment. Such data can then facilitate "evidence-based regulatory reform." A key point about this last approach is that it is collaborative and dialogical, in the sense that regulators, incumbents and new service providers are engaged in an on-going dialogue about the most effective means to gather relevant information and to identify the most appropriate regulatory model.

In order to better understand, the effects, risks and opportunities associated with these regulatory choices, we conducted a simple empirical study of regulatory responses to Fintech in twelve jurisdictions.

In particular, we looked at first-time "venture capital" investments in Fintech companies. The intention was to see whether there was a meaningful connection between levels of investment and the regulatory choice reactive or proactive. When we look at the results of Year-on Year %-growth of first-time venture capital backed companies we get the following Fig. 4. In many cases, this data confirms anecdotal evidence of a slow-down of interest in Fintech. But interestingly, in six of the twelve jurisdictions, there was an increase in investment activity in 2016.

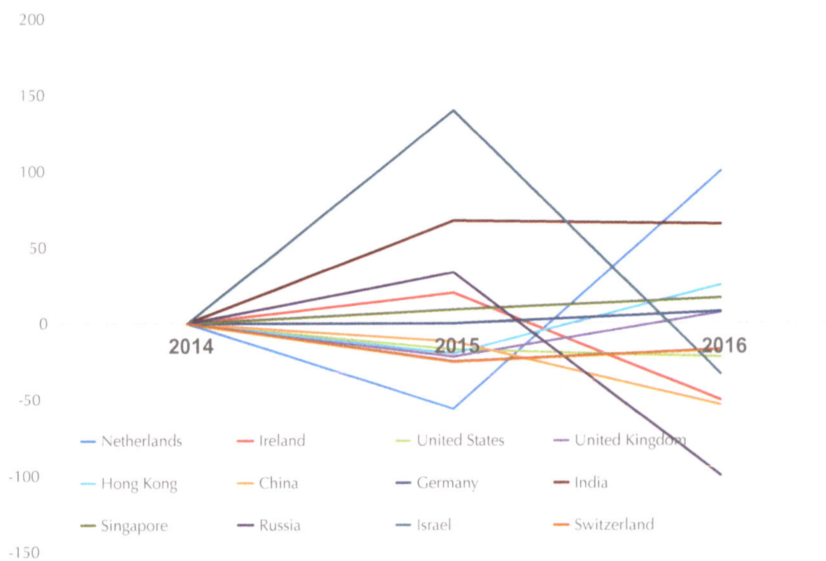

Fig. 4 Year-on-year % growth of "first-time" venture capital-backed Fintech companies (by country, 2014–16). *Source* Pitchbook

The question this data raises is whether there are any signals as to a correlation between more proactive regulatory initiatives and increased activity in the Fintech sector? In those countries in which the response was reactive (red line), there seems to be some evidence of a slowdown. In contrast, in those countries with a more proactive response—particularly involving regulatory guidance (green "dashed" line) or regulatory experimentation (solid green line)—there is some evidence to suggest that this proactive approach makes the jurisdiction more attractive as a potential location for starting Fintech operations (see Fig. 5).

The above analysis suggests in a preliminary way that the regulatory environment does affect the degree of investment and—perhaps as importantly—the willingness of companies to start operations in one jurisdiction, rather than another. In this respect, "regulation matters." This is not to underestimate difficulties of finding an appropriate regulatory regime:

> One obstacle to this goal is that new technologies are often met with highly polarized debates over how to manage their development, use and regulation. Prominent examples include nuclear energy and genetically modified foods…These technologies are characterized by a rapid pace of development, a multitude of applications, manifestations and actors, pervasive uncertainties about risks, benefits and future directions, and demands for oversight ranging from potential health and environmental risks to broader social and ethical concerns. Given this complexity, no single regulatory agency, or even group of agencies, can regulate any of these emerging technologies effectively and comprehensively.[12]

Fig. 5 Year-on-year % growth of "first-time" venture capital-backed Fintech companies (by regulatory approach, 2014–16). *Source* Pitchbook

[12]Mandel (2013), pp. 45 and 136; see, generally, Marchant and Wallach (2013).

It is precisely for this reason that the more dynamic and experimental approach associated with "sandboxes" seems so promising.

4 Innovation "Ecosystems"

Recognizing the importance of more responsive forms of regulation in the context of disruptive technologies is only part of the story, however. We also need to acknowledge that there are other considerations that make a particular ecosystem attractive for Fintech or other AI-related industries. This suggestion points us towards discussion of "innovation systems" and why it is that particular regions become the focus of innovative activity.

4.1 Replicating Silicon Valley

Research has consistently shown that over the last three decades Silicon Valley has been the place to go for anyone interested in setting up a new business, particularly a tech-related business.[13] It has consistently ranked as the best location for launching a new business with global aspirations. Silicon Valley attracts the most funding, it is the most connected and it offers the most opportunities for both innovators and entrepreneurs. Silicon Valley has represented the best bet for anyone with serious aspirations of creating a global business in high growth sectors of the economy.

As a result of this success, policymakers have been drawn to the idea of recreating the success of Silicon Valley in other parts of the world.[14] Initially, this discussion focused on strategies for promoting investment. Generally, this involved two primary types of public intervention into the venture capital market, namely (i) by regulation or de-regulation, which has impact either on the supply side (e.g., on venture capital firms) or on the demand side (e.g., on start-ups) or by (ii) direct public investment schemes.[15] While regulation-de-regulation aimed at creating an enabling environment for private actors to develop their activities, the direct intervention in form of investment schemes effectively enables governmental agencies to fund start-ups in particular sectors and influence investor's behavior.

Yet, clearly more is needed than investment promotion. There is now a much greater awareness of how the success of Silicon Valley is more than just about investment. There is an enormous literature that aims to provide a better understanding of what is needed, in particular seeking to identify the "ingredients" of a successful ecosystem such as Silicon Valley. The aim of developing this understanding is to recreate such an environment elsewhere.

---

[13]See Fenwick and Vermeulen (2015b).

[14]See, e.g., Hwang and Horowitt (2012).

[15]See, e.g., Lerner (2002).

Take Victor Hwang and Greg Horowitt's metaphor of the "rainforest."[16] In identifying the factors necessary to replicate Silicon Valley, Hwang and Horowitt emphasize the importance of a culture in which uncontrolled interactions routinely occur between talent, capital, ideas, and opportunities, i.e., the essential elements in any successful innovation ecosystem. In this type of account, innovation is an unplanned and spontaneous event a feature of the ecology of a rainforest—that is contrasted with the "planned production" of an industrial economy.

Or, Brad Feld. His book, Start-up Communities outlines what he calls the "Boulder Thesis," the elements that he feels have been key to the success of Boulder's start-up ecosystem.[17] Most importantly this means that being led by entrepreneurs. To be successful, an ecosystem must be led by the entrepreneurs themselves, not other players such as governments, universities, or investors. A second factor is long-term commitment: ecosystem builders in the community should take a long-term view, in the order of 20 years or more. Finally, Feld points to a "philosophy of inclusiveness": the ecosystem must be open and welcoming of all. The ecosystem should have regular activities that engage both new and experienced entrepreneurs, as well as investors, mentors, and more.

Or, Steve Case, Co-founder of AOL and author of The Third Wave: An Entrepreneur's Vision of the Future. He has been behind the platform celebrating and investing in emerging start-up ecosystems, the Rise of the Rest movement. In the Rise of the Rest 2018 Ecosystem Playbook, he refers to The Seven Spokes of a Start-up "Hub" seven entities that help to fuel the rise of start-up ecosystems: Local government, Universities, Investors, Start-up support organizations, Corporations, Local media, Start-ups themselves.[18]

But in order to develop a better understanding of what's needed it is also important not to generalize the issue (i.e., to focus on replicating Silicon Valley in some general sense), but also to adopt a perspective that involves looking at how ecosystems might be developed in the context of specific industries or sectors of the economy. Here we would like to suggest that, in the context of AI, a strategy building the ecosystem around established corporations might be particularly effective. In particular, we would like to suggest that whilst regulators have often focused on strategies that aim to develop ecosystems by promoting investment, they tend to neglect the role of partnerships between startups and incumbent, established corporates. Moreover, it is argued that such partnerships are particularly important in the context of "blue sky fields" such as AI.

[16]Hwang and Horowitt (2012), p. 10.

[17]Feld (2016).

[18]Case (2017); for a similar argument, see Fenwick and Vermeulen (2016).

4.2 Building the "Right Kind" of AI Ecosystem I: The EU Experience

The question that therefore needs to be asked is: How to Build the "Right Kind" of AI Ecosystem? National and local governments all now see start-up ecosystems as a necessity for preparing for the future. Putting in place the necessary infrastructure to stimulate the creation, growth and scaling of new and innovative business is now seen as an important and legitimate policy objective for all levels of government.

So, what can governments do to build an effective innovation ecosystem? Traditionally, the focus of policymakers looking to create an innovation ecosystem has been on making more risk capital available for start-up and scale-up companies.

Take Europe as an example. Recently, we can see a steady decline in venture capital activity at all the stages of a start-up's development.[19] This tracks the worldwide trend (according to the data provider and analyst Pitchbook). What appears to be even more worrying is that right now the activity in first-time venture investments is the lowest for over seven years.

In order to stimulate venture capital investments, governments have used several strategies. First, there is long-standing evidence that government support has played a vital role in encouraging entrepreneurship and the launch of start-up companies. For example, governments, in their efforts to establish a sustainable ecosystem, have become the main "post-financial crisis" investor in Europe's start-up scene.

Second, governments often introduce schemes that aim to activate private investments. A recent European example is a joint initiative by the European Commission and the European Investment Fund to set up a Pan-European "VC Fund-of-Funds." The investment of 25% of the total fund-size must encourage private investors, particularly, institutional investors, to invest in the next generation of innovative companies.

Third, regulatory measures can be implemented to make venture capital venture capital more accessible to investors. The proposed amendments to the European Venture Capital Fund (EU VECA) and the European Social Entrepreneurship Funds (EU SEF) are intended to give a boost to the venture capital industry in Europe. Also, investors are encouraged to make venture capital investments through fiscal incentives and tax breaks.

But even if these measures significantly increase the amount of venture capital available, entrepreneurs are not always better off. As is the case with any industry that enjoys a boom, non-specialists will emerge looking to get a piece of the growing pie. There are multiple examples of "new" venture capital investors that have started to invest in innovative companies without doing their proper homework or understanding the rules of the game.

The fear of giving up equity and losing ownership and, eventually, control to less than stellar venture capital investors only feeds a growing skepticism among entrepreneurs about attractive venture capital or other sources of risk-capital. This

[19] See Vermeulen (2018).

means that entrepreneurs prefer bootstrapping, perhaps supplemented with government grants or private loans from family, friends and fools (the "3 Fs").

Although in some cases this can be an effective model, it undoubtedly exposes founder-entrepreneurs to a much greater degree of financial risk and uncertainty. Grants can fill this investment gap but drafting and submitting proposals can take a long time. There is always a lot of competition and managing and administrating the grants can be cumbersome, costly and, ultimately a time-consuming distraction.

As such, venture capital may not represent the missing ingredient for most innovation ecosystems today. The more businesses that are created, the more money becomes available for innovation and innovative firms. There is often an extensive infrastructure supporting entrepreneurs in starting a new business.

Take the example of artificial intelligence. Most of the recently acquired AI companies come from *outside* Silicon Valley (see Fig. 2). Moreover, according to CB Insights, "only" 46% of the acquired AI companies had attracted and received venture capital. So, if it is not really a question of venture capital, what is Silicon Valley doing that is missing in Europe? And what can we do to ensure the success of a sector-focused innovation ecosystem, such as an AI-oriented community of innovators?

4.3 Building the "Right Kind" of AI Ecosystem II: The Role of "Incumbents"

A big part of the solution involves tapping into the experience and know-how of established enterprises. For instance, large, global established corporations often recognize that they must engage with AI, robotics and automation. They have the motive and resources to play a crucial role. And yet, all too often, existing corporate culture and governance structures mean that older, established firms struggle to adjust to new realities. Twentieth-century companies rely too heavily on hierarchical, formal and closed organizations. As such, they are ill-prepared to make the bold and agile business decisions necessary to succeed in a world of constant disruptive innovation.

To survive it is, therefore, imperative for established firms to re-invent their own innovation strategies. This means understanding how to organize for innovation, building and improving on the valuable lessons from the Silicon Valley experience. Crucially, younger firms in the innovation sector are typically organized around the kind of governance principles that provide them with the energy and ideas to constantly innovate, namely a "flat" organization, "open communication" and a "best-idea-wins-culture."[20]

Since these governance principles are more likely to be found in the organization of start-up companies, the "smartest" large corporations try to gain access to this—what Elon Musk has termed the "Silicon Valley operating system"—by cultivating open and inclusive partnerships with entrepreneurs, founders and start-ups

[20]See Fenwick and Vermeulen (2015a).

in the innovation space. When multiple established corporations build relationships of this kind, the basis for a flourishing ecosystem can be put in place. But to build a network, community or ecosystem around this new type of partnership, two processes need to be better understood and embraced. In this way, large corporations can become the crucial link in building innovation ecosystems.

Most obviously, such linkage can drive the kind of genuine opportunities for serendipity highlighted by Hwang and Horrowit discussed above. The first strategy is for established enterprises to use corporate incubator and accelerator programs to put in place an open architecture that offers opportunities for mutual learning. Start-up founder and employees can then get to routinely mix with corporate employees. Such programs have become popular in recent years. In 2017, Amazon, Apple, Facebook, General Electric and Telefónica all announced the opening of new accelerator programs in France, India, the United Kingdom and the United States (see Fig. 6).

The initial and mutual advantages of such an approach seem obvious:

i. *For Established Corporations*. Corporate incubator and accelerator programs allow large firms to engage alongside start-ups and their founders. Such collaborations give them access to ideas and strategies they would never be able to nurture internally.
ii. *For Start-up Companies*. Corporate incubator and accelerator programs are particularly interesting if there is a good cultural match. They can provide start-ups with the necessary capital and deliver tremendous resources in the form of access to relevant knowledge and established international distribution channels.

Nevertheless, the success in any given case will always depend on the structure of the specific program. For instance, there are programs powered by external incubator-

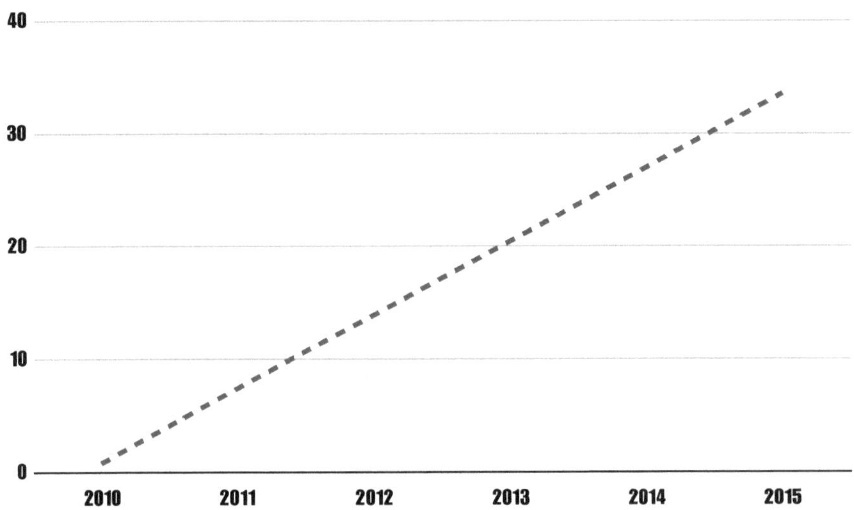

Fig. 6 Launch of corporate incubators/accelerators (2010–15). *Source* Corporate accelerator DB/TechCrunch

accelerator service providers, such as TechStars and Plug and Play. But the clear majority of programs are directly set up by the corporate hosts themselves (see Fig. 4). This would indicate that innovation is more than just the process of involving start-ups.

As Moore's Law begins to slow down, clever high-tech companies are putting more emphasis on the longer-term strategies. In this respect, innovation is not, and never can be, a department. It is a culture that needs to permeate the entire enterprise. This means accepting that innovation cannot be ordered as a product. Without the right environment of clever, motivated, collaborative people, the best ideas will wither and die. Building and then sharing a vision of the domain is crucial. By putting actors together in the right way, the boundaries between corporate and start-up can be blurred, creating new opportunities for positive encounters and interaction.

The second set of strategies essential for building a successful AI ecosystem is to keep things simple and transparent. There can be no misalignment between the interests of the start-ups within the ecosystem and the interests of the corporation (Figs. 7 and 8).

Once a startup has been accepted into the program, the main focus of the corporation should be on assisting and accelerating startups. That means actively connecting them to potentially interesting networks, customers, etc. The possible strategic returns for the corporate partners are a secondary effect or by-product of collaborating with the startups.

Fig. 7 Structure of corporate incubators/accelerators (2010–16). *Source* Corporate accelerator DB/TechCrunch

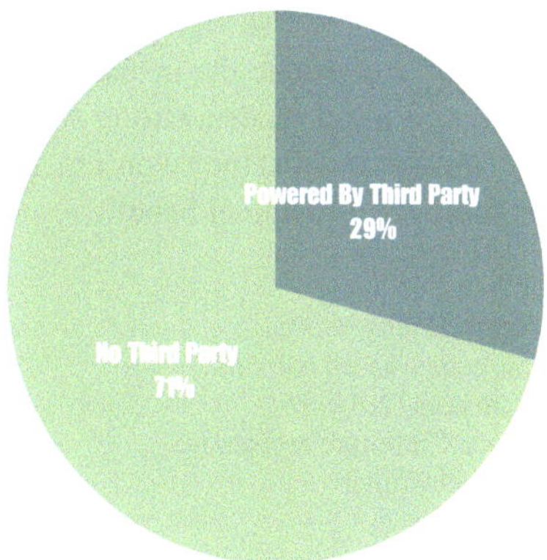

Fig. 8 Support offered by corporate incubators/accelerators (2010–16). *Source* Corporate accelerator DB/TechCrunch

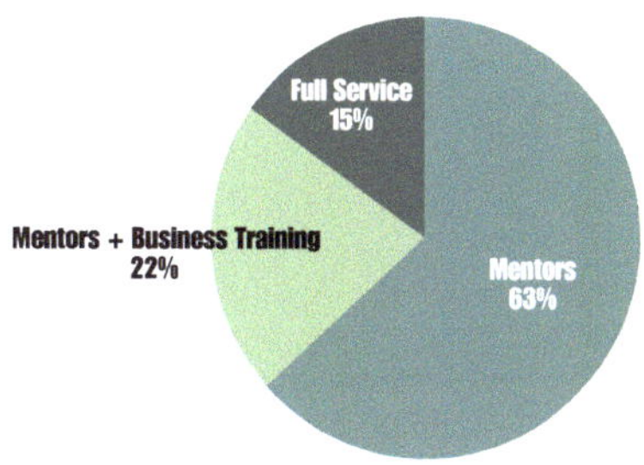

Moreover, by working "cheek-by-jowl" with startup founders, the corporate employees are better able to identify "out-of-the-box" solutions to specific business challenges. The potential benefits are again particularly high in a "blue sky" field such as AI where the technology is complex and "solutions" are not immediately obvious.

From this perspective, non-equity programs seem preferable. Avoiding ownership minority stakes can help simplify relationships, especially at a stage where neither party has a real insight into the true market value of the startup (a common phenomenon in a "blue sky" context). A corporate program that has a first (or even only) objective of making money makes the mistake of trying to execute a business model before the startup has verified that there is one. Such an approach also provides for an important ingredient in developing mutual trust. The absence of direct financial interest in participating startups can help to convince founders that the corporation will not try to appropriate their technology or limit their future options for external financing and strategic partnerships with other corporations (see Fig. 6).

For instance, Microsoft accelerator does not take an equity stake in participating startups and at the same time does not require applicants' products or technology to directly complement or fit with products of Microsoft.[21] To a certain extent, Microsoft relies on serendipity occurrences in the development and future use of technology and cooperation, which may not be always foreseen or obvious (Fig. 9).

Managing minority stakes in portfolio startup companies is often daunting from a legal and accounting point of view. Many startups fear that accepting direct or indirect investments from a corporation will restrict their future funding opportunities and bring about the risk of "negative signaling" should the corporation decide not to support or continue the investment in the future. In this respect, it is important that the corporation sends a strong signal to the ecosystem that they are a trusted partner for startups and that they won't sacrifice a founder-entrepreneur for their own strategic or short-term financial benefits.

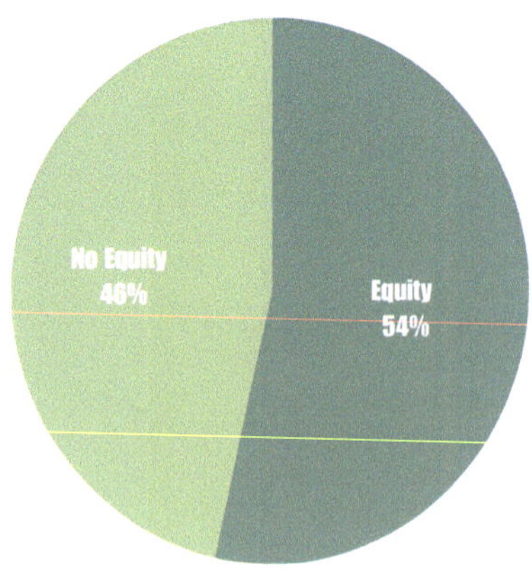

Fig. 9 Equity versus non-equity based corporate incubators (2010–16). *Source* Corporate accelerator DB/TechCrunch

[21]Interview with Maya Grossman, Head of Global Communications, Microsoft Accelerator (Tilburg-Tel Aviv, December 7, 2016).

The most active corporations in the technology sector already understand their role very well. Even after they have acquired a startup, they seek to preserve that startup's unique identity (often by retaining the founders on CEO positions) and do not seek to assimilate it (which has been the conventional wisdom in M&A practice until recently). Moreover, it is precisely this kind of open and inclusive partnering that needs to be at the center of an innovation ecosystem, if it is to be effective. Although large corporations play an increasingly important role in startups ecosystems, policymakers have often been unwilling to recognize this fact.

Does that mean that policymakers are looking in the wrong direction? Not necessarily. But their view does not always offer the full picture. In the context of startups, intensive governmental support has often focused mainly on developing financial markets and catalyzing the venture capital industry. This, oftentimes, prevents regulators from identifying incumbent and successful corporates as crucial "ingredient" of all startup ecosystems.

Only recently several initiatives emerged that aim to support closer startup-corporate cooperation. For instance, the government-supported COSTA (Corporates and start-ups) initiative in the Netherlands, that attracted such corporate giants as Philips, KLM, Unilever and AkzoNobel, promotes a more intensive alliance between corporates and start-ups.[22]

While such initiatives can highlight the importance of startup-corporate cooperation, there is indeed much greater space for policymakers to introduce more measures and thus strengthen the so-called triple-helix cooperation.

5 Conclusion

A well-run innovation ecosystem provides multiple benefits for society in general. It creates the necessary links between complementary sources of risk finance and entrepreneurs. Also, it helps build the capacity of entrepreneurs to identify future partners that are best suited to deliver a meaningful, long-term relationship and give a young firm the best chance of developing its product and scaling successfully. Finally, such an ecosystem helps policymakers develop the "know-how" to implement more dynamic and responsive forms of regulation. In the "right kind" of ecosystem environment, flexible and inclusive processes benefit startups and established companies, regulators, experts and the public.

The ultimate goal is to prepare local and regional ecosystems for a world with a very different level of automation and artificial intelligence. Having a working strategy gives them a credible chance of competing with Silicon Valley. This chapter has identified two promising elements of such an ecosystem that are particularly relevant in the context of AI-related technologies, namely regulatory sandboxes and partnerships between AI startups and incumbent corporations. Such an approach seems to bring clear and tangible benefits to all major stakeholders in such innovation

[22]Costa Program (2016).

ecosystems and has the potential to deliver significant benefits to the community at large.

References

Arbesman, S. (2016). *Overcomplicated: Technology at the limits of comprehension*. New York: Current.

Arner, D., Barberis, J. N., & Buckley, R. P. (2016a). The evolution of Fintech: A new post-crisis paradigm? UNSW Law Research Paper No. 62.

Arner, D., Barberis, J. N., & Buckley, R. P. (2016b). FinTech, RegTech and the reconceptualization of financial regulation. University of Hong Kong Faculty of Law Research Paper No. 35.

Beck, U. (1992). *Risk society: Towards a new modernity*. London: Sage.

Bennett Moses, L. (2011). Agents of change: How the law 'copes' with technological change. *Griffith Law Review, 20,* 763–794.

Black, J., Hopper, M., & Band, C. (2007). Making a success of principles-based regulation. *Law & Financial Markets Review, 1,* 191–206.

Black, J. (2009). Forms and paradoxes of principles-based regulation. Available via LSE Law, Society and Economy Working Papers: http://eprints.lse.ac.uk/23103/1/WPS2008-13.pdf/. Accessed 1 May 2018.

Calliess, G. P., & Zumbansen, P. (2010). *Rough consensus & running code: A theory of transnational private law*. London: Hart.

Callon, M., Lasoumes, P., & Barthe, Y. (2009). *Acting in an uncertain world: An essay on technical democracy*. Cambridge: MIT Press.

Case, S. (2017). *The third wave: An entrepreneur's vision of the future*. New York: Simon & Schuster.

Costa Program. (2016). The COSTA programme: Bringing corporates and start-ups together. http://www.brainport.nl/en/news-developments/costa-moet-start-ups-en-bedrijven-verbinden. Accessed 1 May 2018.

Feld, B. (2016). *Startup communities: Building an entrepreneurial ecosystem in your city*. New York: Wiley.

Fenwick, M., & Vermeulen, E. P. M. (2015a). The new firm: Staying relevant, unique & competitive. *European Business Organization Law Review, 16,* 595–623.

Fenwick, M., & Vermeulen, E. P. M. (2015b). Alternatives to silicon valley: Building global business anywhere. *EUIJ-Kyushu Review, 5,* 27–68.

Fenwick, M., & Vermeulen, E. P. M. (2016). Global startup communities. In Stichting Maatschappij en Onderneming. http://smo.nl/global-start-up-communities/. Accessed 1 May 2018.

Fenwick M, Vermeulen EPM (2017) How to respond to artificial intelligence in Fintech. *Japan Spotlight,* July/August, 16–20.

Fenwick, M., McCahery, J. A., & Vermeulen, E. P. M. (2018). Fintech and the financing of SMEs and entrepreneurs: From crowdfunding to marketplace lending. In D. Cumming & L. Hornuf (Eds.), *The economics of crowdfunding: Startups, portals & investor behavior*. London: Palgrave Macmillan.

Financial Conduct Authority. (2015). Regulatory sandbox. Available via UK Financial Conduct Authority https://www.fca.org.uk/publication/research/regulatory-sandbox.pdf/. Accessed 1 May 2018.

Haddad, C., & Hornuff, L. (2016). The emergence of the Global Fintech Market: Economic and technological determinants. CESifo Working Paper Series No. 6131.

Hwang, V. W., & Horowitt, G. (2012). *The rainforest*. Los Alto Hills: Regenweld.

JASON. (2017). Artificial Intelligence for Health & Health Care https://www.healthit.gov/sites/default/files/jsr-17-task002_aiforhealthandhealthcare12122017.pdf. Accessed 1 May 2018.

Kaal, W. A. (2013). Dynamic regulation of the financial services industry. *Wake Forest Law Review, 48,* 791–828.

Kaal, W. A. (2014). Evolution of law: Dynamic regulation in a new institutional economics framework. In W. A. Kaal, M. Schmidt, & A. Schartz (Eds.), *Festschrift Zu Ehren von Christian Kirchner*. Tübingen: Mohr Siebeck.

Lerner, L. (2002). When bureaucrats meet entrepreneurs: The design of effective public venture capital schemes. *The Economic Journal, 112,* 73–97.

Mandel, G. N. (2013). Emerging technology governance. In G. E. Marchant, K. W. Abbot, & B. Allenby (Eds.), *Innovative governance models for emerging technologies*. London: Edward Elgar.

Marchant, G. W., & Wallach, W. (2013). Governing the governance of emerging technologies. In G. E. Marchant, K. W. Abbot, & B. Allenby (Eds.), *Innovative governance models for emerging technologies*. London: Edward Elgar.

MIT Sloan School of Management. (2016). Fintech platforms & strategy. MIT Sloan Research Paper No. 5183-16.

Sachs, S. N. (2011). Rescuing the strong precautionary principle from its critics. *University Illinois Law Review,* 1285.

Sunstein, C. R. (2005). The precautionary principle as a basis for decision making. *Economists' Voice, 2,* 1–9.

Vermeulen, E. P. M. (2018). Capital markets union: Why venture capital is not the answer to Europe's innovation challenge. In D. Busch, G. Ferrarini, & E. Avgouleas (Eds.), *Capital markets union in Europe*. Oxford: Oxford University Press.

The Rise and Regulation of Drones: Are We Embracing Minority Report or WALL-E?

Pam Storr and Christine Storr

Abstract The popularity of drones has increased exponentially over the last few years. The advance of technology has not only led to drones being used by a greater number of people, but also allowed for the technological capacity of drones to increase at a great pace. How drone technology can be used, and by whom, has in turn led to regulatory concerns. Is the current legal framework equipped for such technology and is it adapting with necessary changes at sufficient speed? The ways in which drones can, and have already, been used and misused have highlighted the need for certain regulatory development. This chapter will deal with legal issues connected to drones from a European perspective. It focuses on laws that are affected by the use of drones, specifically within the areas of surveillance, privacy and aviation. The underlying theme of the chapter is whether the developing legal framework manages to deal with the challenges surrounding the rapid rise in drone usage. The future of drones and the potential consequences of the legal framework being adopted will also be addressed.

Keywords Drones · Surveillance · Privacy · Data protection · Aviation

1 Introduction

Drones come in many shapes and forms. Traditionally a tool used by the military, drones are now being used by a wide range of actors within both the public and private sectors. These actors, along with individuals, are making growing use of drone technology due to the falling cost of drones and the benefits they can provide. The ways in which drone technology is being deployed, and the variety of actors involved in such deployment, naturally raise many questions from a legal perspective.

P. Storr (✉)
Legal Consultant and Teacher in IT law, Stockholm, Sweden
e-mail: pam@legalgeekgirl.com

C. Storr
Faculty of Law, Stockholm University, Stockholm, Sweden

M. Corrales et al. (eds.), *Robotics, AI and the Future of Law*, Perspectives in Law, Business and Innovation, https://doi.org/10.1007/978-981-13-2874-9_5

Although drones can be used for a wide variety of purposes, from tracking a crime suspect to surveying the site of a natural disaster, their primary feature is the collection of data. The type and amount of data collected and its subsequent use is therefore of particular importance; laws dealing with surveillance and privacy rights must therefore be addressed in relation to drones. Additionally, due to their differing size and capacity drones have the ability to fly almost everywhere, only limited by the ability to take-off and land, size or battery life; where drones are able to fly, and where such flight should be permitted, therefore bring aerial regulations into play.

The extent to which drones are covered by the existing legal framework within these three areas: surveillance, privacy and aviation, therefore forms the basis of this chapter.

2 The Rise of Drones

The term "drone" can refer to many different objects. It is the word most commonly used when referring to an unmanned aerial vehicle, or UAV for short; a multitude of alternative terms exist, such as remotely piloted aircraft or vehicle (RPA and RPV respectively), remotely operated aircraft (ROA) or, simply, unmanned aircraft (UA). Even the term model aircraft (MA) is used when referring to drones used by individual hobbyists. The term "drone" will be used in this chapter, as this is the standardized and seemingly all-encompassing term for such objects. However, it is clear that the regulation of drones requires a more nuanced approach in its terminology, due to the differing kinds of drones, their capabilities, size, weight and other relevant attributes.[1]

Take a military drone, which can be comparable in shape and size to a small, traditionally piloted aircraft. How the drone is able to take off and land, its cruising altitude, flying range, and on-board technology are all affected by its physical dimensions. A small drone with rotating blades in a circular pattern, would take off and land in a fashion similar to a helicopter, fly at a lower altitude and have less on-board technology compared to the larger drone. How a drone functions and its effects on the surrounding environment will therefore depend greatly on the particular drone's physical attributes. These types of differences must therefore be considered from a regulatory perspective, and this is before we even consider the ways in which a particular actor wishes to use a drone.[2]

[1]For a more detailed discussion of terminology and drone categories, see Clarke (2014b), pp. 234–236.

[2]For more information on different types of drones and their uses, see Sandbrook (2015), pp. 636–638.

2.1 Drone Use

Drones are already being used on a fairly large scale, performing different kinds of functions and are popular within a number of different sectors. *Delivery* by drone is an emerging sector, with companies such as Swiss Post,[3] Deutsche Post DHL[4] and Amazon[5] developing delivery systems for delivering letters, parcels and documents by drone.[6] Delivery of food and medical supplies, although not yet widely used, would be a natural development of drone delivery.

Another popular use of drones is *photography* and *aerial surveillance*. Drones are ideal for taking aerial shots of buildings and land, for example, a house that is being put up for sale, or examining the consequences of a natural disaster. Aerial filming can include undertaking search and rescue missions, or monitoring large areas of agriculture. For law enforcement purposes, drones can, for example, be used to monitor crowds, control borders and track criminal suspects.

Although the majority of drone uses are carried out by commercial actors and the public sector, *hobbyists* make up another emerging sector. Smaller drones are now being sold and are widely accessible to individual consumers.

As the technology develops so do the potential uses of drones, the types of data that can be collected and the subsequent analyses that can be performed. The collection of data through attaching additional sensors on drones can, for example, be used to monitor pollution levels in a particular area.

The popularity of drones varies greatly from country to country, likely due to factors such as the economy, technological development and availability of research facilities. Within Europe, drone development has been prevalent in the UK; Amazon has, for example, carried out testing of drone delivery there after regulatory problems the company faced in the US.[7] As mentioned above, Germany and Switzerland are also examples of early drone deployment within the EU.

2.2 Drone Misuse

As drone usage has increased in recent years, the potential for drone misuse has also been demonstrated. As with any mode of transport, *dangerous operation* can lead to damage to persons or property; without proper control and response drones can

---

[3]Swiss Post, Drones in logistics: Transporting goods by airmail, https://www.post.ch/en/about-us/company/innovation/swiss-post-s-innovations-for-you/drones-in-logistics. Accessed 4 February 2018.

[4]Deutsche Post DHL Group (2016).

[5]Amazon, Amazon Prime Air, https://www.amazon.com/Amazon-Prime-Air/b?node=8037720011. Accessed 4 February 2018.

[6]See, further, Dorling et al. (2017), p. 70.

[7]Luppicini and So (2016), p. 115.

cause collisions mid-air, on the ground or fly into structures.[8] Even if no physical damage occurs, dangerous operation has other negative consequences; flying drones in no-fly zones, such as near airports, can lead to disruption to air travel amongst other things.

Through their *collection of data*, drones also have the potential to be misused in other ways; examples are collecting data on an individual, organization or body without their knowledge or consent and without a legal basis, along with unintentional data collection, where data other than that originally intended is collected. As drones become more sophisticated, the potential threat of such collection and misuse also increases.

External factors are also a threat needing to be addressed, in the form of *physical or technological interference*. Interference of a physical nature would, for example, include attempting to shoot down a drone, perhaps leading to injury, loss of data and so forth. Technological interference is generally carried out from a location further afield, and could include interception of drone communications and tampering with drone data.[9]

The misuse of drones is likely to continue and increase in frequency as the technology becomes more widespread and affordable. In this regard it is important from a regulatory perspective to consider not only *actual* misuse, but also the *potential* for and the *perceived* misuse of drones. Even where misuse does not occur in practice to a large extent, the belief that drones are used as a surveillance tool, by, for example, the state and law enforcement authorities, as in science fiction films such as Minority Report, is a problem in itself. Where citizens believe their rights, including the right to privacy and personal data protection, are unjustly restricted they are more likely to hold a negative view of the particular technology in question.

2.3 A Legislative Framework for Drones

Minimizing the risks of drone misuse is one task for legislators when overseeing the current legal framework. Drones can be used in numerous ways, and by a wide variety of actors, as detailed above. The main challenge from a legislative perspective therefore stems from the complexity of the subject: new technology that is continuing to develop, multiple potential uses (and misuses), and users from both the public and private sectors, along with individuals. As a result, a number of different areas of law need to be considered from a number of different perspectives when assessing whether the current legal framework adequately deals with the rapid rise in drone usage.

Failing to address these issues from a regulatory standpoint would, however, threaten to limit the potential of drones. In addition to the concerns of citizens mentioned above, private actors may be more reluctant to invest in drone technology if

[8]See, e.g., Shelley (2016), pp. 73.74 and Cracknell (2017), p. 3055.

[9]For further, specific examples of the misuse of drones, see Rule (2015), p. 157.

the legal consequences of drone usage were unclear. Future development could be at risk without legal guidance on matters such as *where* drones are permitted to fly and *who* has responsibility for drone operations. Innovative and beneficial uses of drones could therefore be stifled due to an insufficient regulatory regime.

Drone technology is here, and arguably here to stay. Assessment of the legal environment for drones is clearly necessary to avoid misuse of the technology, either due to a lack of regulation or loopholes in existing regulations, but also to ensure a coherent legislative framework and encourage the use of drones for society's overall benefit. If done correctly, this framework has the possibility to allow for the technology to be used in innovative and, perhaps as yet, undiscovered ways.

3 The Regulation of Drones

Drone regulation will be tackled from a European Union (EU) perspective, considering the current overarching legal framework. Where necessary, comparison will be made with specific countries' regulatory systems and the challenges that other countries face in drone regulation. While several regulatory aspects could be discussed at length, the focus will be on the most prominent areas of regulation, namely surveillance, privacy and aviation.

3.1 Surveillance

Surveillance of citizens is to an extent regulated by EU law. Directive 2002/58/EC of the European Parliament and of the Council of 12 July 2002 concerning the processing of personal data and the protection of privacy in the electronic communications sector (Directive on privacy and electronic communications) ensures confidentiality of communications and prohibits surveillance or interception of communications and traffic data. Exceptions to this prohibition are only granted where an individual has given consent or where a legal authorization exists and specific requirements are fulfilled.[10] The Directive provides a framework on surveillance, and requires Member States to ensure confidentiality of communications through their national law. However, the main competence within the field of surveillance is left to the individual Member States, as part of their national criminal law regimes.

When considering surveillance by drone, a comparison can be made with wiretapping or electronic surveillance laws, as in both cases surveillance occurs from a distance and subjects are generally not aware that their communications or movements are being tracked. Whether drone surveillance would be covered by electronic surveillance laws, however, depends on the particular national legislation. The Directive on privacy and electronic communications assumes that a publicly available

---

[10] Article 5 Directive 2002/58/EC.

electronic communications service is being used, which is not necessarily the case when carrying out drone surveillance.[11] As such, drone surveillance may well fall outside the scope of existing surveillance regulations.

Where this is the case, protection of individuals is provided through other means. The Charter of Fundamental Rights of the European Union provides a broad protection for individuals, encompassing private and family life, home and communications, along with protection of personal data.[12] These rights can, in accordance with principles of EU law, only be restricted where necessary and proportionate.[13] Such restriction has been accepted in the case of law enforcement activity, but would, for example, not be possible for the furthering of a corporation's interests.

Generally speaking, surveillance laws have been developed that balance the power of law enforcement and the need for public safety against the protection of fundamental individual rights. Provided these laws are followed and applied in a commonsense manner, there is no reason why drones used by law enforcement agencies should pose any great risk to this balance. Certain legal aspects would of course need to be addressed and alignment with other national principles may be necessary. Examples include the powers of law enforcement authorities to carry out drone surveillance, whether there exists a prerequisite of a sufficiently serious crime for such surveillance, and under what circumstances a court order is needed before drone surveillance is lawfully undertaken.[14]

Overreach by the state in surveiling their citizens is, of course, a concern. States have a fairly wide discretion in using surveillance measures, particularly for reasons of national security. Such exceptions to the general legal framework are necessary, but clear and robust safeguards are also required. While not a matter purely relevant to drones, surveillance carried out in a questionable manner raises a number of concerns, particularly as it is not known the type of information that is collected, to whom the information relates, nor how such information is subsequently used.[15] Where a lack of transparency and accountability exist, the potential for overreach by the state is a real concern and also leads to questions of an ethical nature.[16]

One issue to consider in relation to surveillance by drone is that, unlike wiretapping and traditional forms of communications surveillance, information on a vast number of individuals besides the target of a particular law enforcement operation is likely to be captured. Rather than focusing on a specific individual's communications, surveillance by drone can instead focus on a number of different aspects, of which communications is just one; movements of an individual and the filming of property or of a specific area are just some examples. How this surveillance is used in practice

[11] See Article 2 (d) Directive 2002/58/EC.

[12] Articles 7 and 8 Charter of Fundamental Rights of the European Union.

[13] See also Article 52 Charter of Fundamental Rights of the European Union.

[14] In this regard, recent discussions regarding data retention and the power of law enforcement authorities to access retained data are pertinently relevant.

[15] For example, a secret drone program existed in the US for a number of years, see Weismann (2015).

[16] For an analysis of these issues, see West and Bowman (2016).

must therefore be carefully considered, particularly in terms of individual privacy rights.[17]

Another area of relevance at a national level is surveillance by private actors. Surveillance laws naturally focus on the state, detailing the types of activity law enforcement authorities are permitted to carry out. With the introduction of electronic communication surveillance, the role of telecommunication providers has been addressed through legislation such as the Directive on privacy and electronic communications, detailing their duties towards both the state and the individual. Other private actors have, however, traditionally not been the focus of surveillance regulation, as they generally have been unable to perform surveillance due to a lack of technology and resources. This issue is somewhat complicated as technology develops; some private actors may be just as able as the state to acquire technology equipped with surveillance capabilities. Larger drones capable of flying long distances may be too expensive for the majority of individual users, although perhaps not for a large commercial actor. The impact of such technology is not unique to drones; other types of developing technologies such as smart cars and Internet-of-Things (IoT) applications possess similar surveillance capabilities. Such developments in surveillance technology and the increasing role of private actors lead to the question of whether additional surveillance regulation is required, either at the EU or national level.

Certain regulation of private actors exists, such as prohibitions on hacking or the interception of electronic communications.[18] The focus of such regulation is access to stored data or data being communicated by another party, in other words the electronic equivalent of theft or opening a sealed letter. Surveillance technology, however, poses a new risk, namely the direct collection of data by a private actor. Surveillance by actors other than law enforcement authorities, whether by individuals, companies or even other public bodies is therefore a matter that should be dealt with by legislators before problems or abuses arise. Any such regulation should, for the reasons mentioned above, not specifically focus on drones, but rather adopt a more general, encompassing approach in order to deal with the impact of new technology at large.

When considering new types of surveillance, some countries will be more prepared than others to deal with developments in the use of technology. The adoption of CCTV, for example, has already led to regulatory discussion on surveillance and legislative amendments. The most notable example in this context is the UK, where CCTV use is prevalent in most major cities, and has been for a number of years. A surveillance camera code of practice was published in 2013 and is obligatory for public bodies and voluntary for private actors[19]; a Surveillance Camera Commissioner was also established as a result of the Protection of Freedoms Act 2012, although lacking enforcement or inspection powers.[20] In countries where discussions on surveillance technology have taken place in recent years, the regulatory frame-

[17]See, below, Sect. 3.2.

[18]See, e.g., Council of Europe's Convention on Cybercrime ETS No. 185.

[19]Home Office (2013).

[20]See Surveillance Camera Commissioner, https://www.gov.uk/government/organisations/surveillance-camera-commissioner/about. Accessed 4 February 2018.

work should be at least somewhat prepared for the introduction of drones and other new surveillance technologies.

Regulatory discussions at the national level are, however, necessary in order to make changes and adaptations allowing for new surveillance technologies such as drones. To illustrate the complexity of surveillance regulation, Sweden has a Camera Surveillance Act (*Kameraövervakningslag*, 2013:460) that was updated in 2017 so as to exclude private actors' usage of "unmanned aircraft."[21] Prior to the change, drone usage by state and private actors alike would fall within the scope of the Act. As a result of a legislative amendment private actors are now exempt from, amongst other things, the requirement to seek permission before surveillance is used in a public place.[22] This means that different rules apply in practice whether drone surveillance is being carried out by (a) an authority, where the Camera Surveillance Act will apply, (b) a private actor (individual or company), where the normal data protection provisions will apply, but only in cases where an individual is able to be identified via the surveillance,[23] and (c) media companies with constitutional protection, where neither the Camera Surveillance Act nor the data protection provisions will be applicable.

There are many possible approaches to such regulation, focusing, for example, on one or more of the following: (i) the action in question, i.e., surveillance and the collection of data; (ii) the technology itself, such as drones; (iii) the type of actor involved, i.e., public sector, private sector or individual. The approach in Sweden has been a combination of these factors, focusing on both the actor involved and the technology, in this case a wider category of cameras in which drones are included. As the Swedish example shows, legislative amendments by EU Member States at the national level risk a piecemeal approach to new surveillance technologies, and disparate rules from country to country.

3.2 Privacy and Data Protection

Privacy is an important aspect in the regulation of drones, due to the amount of data that can be collected as a result of their use. These privacy risks often stem from a lack of transparency according to the Article 29 Data Protection Working Party[24]; for example, people are in many cases unaware that a drone is being used to collect data in a particular area, of what information is being collected, of who is operating a drone, what capabilities it has and for what purpose the data is being used.[25]

Privacy as an individual right has a long history within the EU, and has been in the forefront in recent years due to the new Regulation (EU) 2016/679 of the European

[21] Section 5 a Camera Surveillance Act (2013:460).

[22] Section 8 Camera Surveillance Act (2013:460).

[23] See, further, Sect. 3.2 below.

[24] The Article 29 Data Protection Working Party is an EU advisory body on data protection.

[25] Article 29 Data Protection Working Party (2015), p. 3 and p. 7.

Parliament and of the Council of 27 April 2016 on the protection of natural persons with regard to the processing of personal data and on the free movement of such data, and repealing Directive 95/46/EC (General Data Protection Regulation, GDPR), that came into force in May 2018. As the GDPR provisions apply to all private and public actors processing personal data, its impact will be widely felt. Important to state in this regard is that the GDPR does not apply to law enforcement authorities in the prevention, detection and investigation of crimes or cases where there is a threat to public security[26]; nor does the GDPR apply to purely personal activities.[27] Drone usage by law enforcement will therefore generally fall outside the scope of the rules, however an individual operating a drone in public does not class a purely personal activity and will therefore be subject to the rules.

The GDPR builds on existing data protection legislation and follows the general principles that were established in Directive 95/46/EC of the European Parliament and of the Council of 24 October 1995 on the protection of individuals with regard to the processing of personal data and on the free movement of such data (Data Protection Directive, DPD). Personal data, or data relating directly or indirectly to an individual,[28] shall, for example, be collected for a clear and explicit purpose, and can only subsequently be processed in a manner compatible with that stated purpose.[29] Data collected through the use of drones must adhere to these rules in exactly the same way as data collected through the use of other technologies.

3.2.1 Impact of the GDPR

Where the GDPR could play an important role in relation to drones is in their development and deployment. This is because the GDPR includes in its provisions obligations to use available technology to protect privacy. Data protection impact assessments (DPIAs, also known in some jurisdictions as privacy impact assessments, PIAs) must be carried out, particularly where new technologies are used or there is a systematic monitoring of a publicly accessible area on a large scale.[30] Those intending to use drone technology will therefore need to assess the impact a drone will have, for example, the different sensors being used, what personal data is collected, how this data is to be used and so forth.

In addition, the GDPR requires privacy to be built in from the outset, first by limiting the amount of personal data and subsequently by protecting the data that is processed. Technological and organizational measures are to be implemented in order to ensure the amount of personal data being processed is limited to that necessary;

[26] Article 2 (2) (d) GDPR.

[27] Article 2 (2) (c) GDPR.

[28] Article 2 (a) DPD, Article 4 (1) GDPR.

[29] Article 6 (1) (b) DPD; Article 5 (1) (b) GDPR.

[30] Articles 35 (1) and 35 (3) (c) GDPR.

this requirement is referred to as *data protection by default*,[31] and applies from initial collection through further processing, storage and access.

Technological and organizational measures are also to be implemented to protect the data being processed; this requirement is referred to as *data protection by design* and includes such measures as pseudonymization.[32] When deciding on these measures, the individual circumstances are taken into account, such as the type and amount of data being processed, the cost of potential measures and the available technology. Processing a large amount of personal data, or where there exists a heightened risk to individuals due to the nature of the data being processed, will, for example, lead to a stricter protection requirement.

In order to adhere to the GDPR provisions, drone operators will therefore need to take a number of steps to protect data that directly relates to a particular individual or an individual who can be identified through location data or other factors.[33] Ironically, the further the technology develops, for example, through improved sensors and higher resolution images, the more likely it is that an individual can be identified, leading to additional protection obligations for drone operators and others making use of such technology.

The task of enforcing the GDPR provisions has been placed on the national data protection authorities.[34] An important factor in this regard will be the strength of the relevant authority's enforcement mechanisms, whose resources are likely to be spread thin as a result of their increased role in "policing" the area of data protection and issuing fines in cases of serious violations.[35] Another aspect to consider is whether data protection authorities have sufficient expertise in relation to drones and other new technologies in order to provide an effective oversight role.

3.2.2 Remaining Privacy Challenges

Some of the regulatory gaps noted above in relation to surveillance[36] are therefore filled by data protection legislation, at least where data is or can be connected to an individual. Private actors are subject to data protection legislation and the GDPR increases obligations on organizations that process personal data. However, there still remain a number of privacy challenges linked to the use of drones.

One such challenge is that of *unintended data collection*. A drone taking photographs of a specific object, for example, a house or piece of land, may inadvertently include other objects, such as people, buildings or neighboring land. As such, this unintended data may or may not be personal in nature. Questions arise as to how this data is processed and the procedures in place for its storage and deletion. A seemingly

[31] Article 25 (2) GDPR.

[32] Article 25 (1) GDPR.

[33] See the definition of personal data in Article 4 (1) GDPR.

[34] Article 57 (1) (a) GDPR.

[35] Article 83 GDPR.

[36] See above Sect. 3.1.

obvious step to counter this challenge would be to make use of the technology itself; the technological advances which have allowed for the multitude of sensors to be placed on drones also allow for the protection of privacy. Drones can be programmed to, for instance, blur out unwanted parts of an image. Such an approach has the added benefit of data minimization.[37] Google Maps is one example where image blurring has already been put into use on a large scale, blurring out individuals and other identifiable objects such as car registration plates in "street view."[38] Drones could also be programmed to prevent data collection outside a specific area by disabling certain sensors, making use of geo-fencing technology.[39]

Another, linked, challenge is the *purpose of drone usage*. Where data being collected by a drone is not personal in nature, the data protection framework will not apply. However, just because no personal data is being collected does not mean that a drone is being used for a legitimate purpose. A company flying a drone may be trying to find out what new technologies or procedures are being used by a rival company, or perhaps stumbles across trade secrets that would not have been visible to the public but for an aerial vantage point. An individual may be making use of drone technology to find out where law enforcement officials are stationed or the positioning of crowd control measures in advance of a planned demonstration, in order to evade detection for one reason or another.

Challenges of such nature may require additional regulatory measures, perhaps tailored to particular sectors. Where drones are increasingly commonplace within a certain sector, for example, by estate agents and delivery companies, a specific code of conduct tailored to the sector and its drone usage may be advisable. These codes can be adapted to the sector's specific use of drones, providing a complement to the legal framework and a more practical and user-friendly guide for drone operators. More detail can, for example, be provided on what is acceptable versus non-acceptable usage within a specific sector. A number of such codes have already been developed, drawn up by various organizations as drones have gained popularity.[40] Drone certification programs could also be introduced so that organizations are able to show their commitment to best business practices. Transparency of drone operations is also a way to counteract such challenges, for example, through the release of annual reports detailing how drones are used, an approach adopted by public bodies in the US.[41]

Additional regulation of these kinds should encourage the use of drones within reasonable bounds, where organizations agree to adhere to accepted standards. While not solving all problems, particularly in relation to the example above of an indi-

---

[37] See Sect. 3.2.1 above.

[38] Volovelsky (2014), pp. 319–320 and Sandbrook (2015), p. 644.

[39] See, further, McNeal (2016), pp. 394–395.

[40] For example, at a general level by the Association for Unmanned Vehicle Systems International (AUVSI) and at a sector-specific level by the Professional Society of Drone Journalists (PSDJ). In some cases, codes of conduct have even been written at the state level, for example, in the UK, where drones are included as part of a more general code on surveillance cameras, see Information Commissioner's Office, Guide to Data Protection: CCTV, https://ico.org.uk/for-organisations/guide-to-data-protection/cctv/. Accessed 4 February 2018.

[41] The White House (2015), Section 1 (d). See further McNeal (2016), p. 371.

vidual's motivation, such regulation goes some way towards reducing the risk of improper drone usage.

Another way of dealing with such challenges would be through the introduction of drone registration. A registration requirement would facilitate the procedure of ascertaining the operator of a drone, particularly relevant in situations where the purpose of drone usage is called into question. This would also be a requirement that could be applied to individual drone operators. Combining registration information with flight plans and other logs, comparable to that of air travel, would be a way in which to ensure that any improper usage could be tracked and followed up by the relevant authorities. Regulation through registration also encourages safe drone operation. Safety and aerial regulations have, logically so, been a predominant area of focus of drone regulation thus far and will be dealt with in the next section.

3.3 *Aviation*

Perhaps the most important aspect of drone regulation is aviation safety, both in terms of the technology itself and drone operation. In order for drones to gain widespread acceptance from a policy perspective, the technology itself needs to be reliable and the number of drone incidents kept to a minimum. Product safety standards are a crucial aspect in the manufacture of drones, necessary in order to avoid malfunction and injury.[42] Aviation safety standards are also at the forefront of drone operation safety, in much the same way as for traditional aircraft.

If traditional aircraft for one reason or another cannot fly in a particular area, for example, near to an airport or in a government-controlled area, drones should as a general rule not be permitted either. Regulations may need to be updated so as to include drones in the list of objects that cannot fly in these no-fly or restricted zones. Failing to do so could lead to widespread disruption; airports have in recent years been forced to temporarily suspend aircraft take-offs and landings due to drone use in the vicinity.[43]

The technology in drones has the capability to overcome this challenge by automatically programming in no-fly zones. If a drone reaches a no-fly zone it can either power down in order to land in a safe area or automatically divert its course to avoid its path entering the zone.[44] This type of geo-fencing programming would most probably have to be carried out at an individual level at the current time, but could arguably be programmed by manufacturers if and when the flying zones for drones are agreed upon at a national or international level. A software update to drones already in operation could also be a way to reflect changes in no-fly zones over time.

---

[42]There have been reports of drones "falling" out of the sky; one drone manufacturer even recalled its drones due to loss of power mid-operation, see The Verge (2016).

[43]This disruption has occurred in a number of countries, see, e.g., BBC News (2016), SVT News (2017) and The Independent (2017).

[44]This type of solution already exists today; see Corcoran (2014).

In-built technology can also help to avoid mid-air collisions involving drones. In the same way as a drone can divert from a no-fly zone, it could change its course if a sensor picks up an object that is in its path risking a collision. Drones therefore have the potential to develop into a much safer form of aerial device than traditional aircraft at their outset.[45]

Safety is a primary focus of aviation organizations. In the wake of drones, it has been debated whether aviation organizations need to adapt their regulations, and if so at what level. The International Civil Aviation Organization (ICAO) in its Convention on International Civil Aviation states that the regulation of pilotless aircraft is regulated at the national level.[46] Within the EU, drones weighing less than 150 kg are left to Member State regulation[47]; this should comprise the majority of drones being used today. Only heavier drones, such as those similar in size to small aircraft, fall within the competence of the European Aviation Safety Agency (EASA).

The focus of drone aviation regulation is therefore predominantly national, rather than regional or international, in nature. Other EU legislation may, however, be relevant, such as Council Directive 85/374/EEC of 25 July 1985 on the approximation of the laws, regulations and administrative provisions of the Member States concerning liability for defective products, and Directive 2009/48/EC of the European Parliament and of the Council of 18 June 2009 on the safety of toys.[48] A number of EU countries have introduced drone aviation regulations in recent years, which have included requirements relating to drone registration, weight and flying altitude.[49] The current regulatory approach means in practice that implementation mechanisms and the scope of regulations are decided on a country-by-country basis, resulting in differences in the content of drone aviation regulations between Member States. This fragmented approach has been criticized by the European Commission from a market-growth perspective, arguing that it hinders the development of drone usage within the EU.[50]

However, the need for harmonization of aviation regulations has become apparent in recent years. In 2015 the EASA issued an opinion on the Introduction of a Regu-

---

[45] There have even been recent news reports of passenger drones, see The Guardian (2016).

[46] Article 8 Convention on International Civil Aviation, Doc 7300/9, Ninth Edition, 2006. The ICAO is a UN agency, established in 1944 to manage the administration and governance of the Convention on International Civil Aviation (also known as the Chicago Convention), www.icao.int.

[47] Annex II (i), Regulation No 216/2008 of the European Parliament and of the Council of 20 February 2008 on common rules in the field of civil aviation and establishing a European Aviation Safety Agency, and repealing Council Directive 91/670/EEC, Regulation (EC) No 1592/2002 and Directive 2004/36/EC.

[48] A number of factors are relevant in deciding whether or not a product is to be classed as a toy and therefore fall within the definition of the Directive. It is ultimately a question for Member States in their transposition of the Directive into national law. See further European Commission, Guidance on Toy Safety, https://ec.europa.eu/growth/sectors/toys/safety/guidance_en. Accessed 4 February 2018.

[49] A comprehensive overview of national regulations, for both recreational and professional uses, can be found at www.dronerules.eu, co-funded by the EU.

[50] European Commission (2014).

latory Framework for the Operation of Drones,[51] and with a proposal to regulate the operation of small drones in May 2017.[52] An EU Regulation has also been proposed (proposed Regulation on Civil Aviation),[53] with the primary aim of creating "a Union framework for safe integration of unmanned aircraft into the European airspace."[54] The proposal specifically addresses the safety of drones and takes a broad perspective, covering the design, production, operation and maintenance of drones.[55] Rather than categorizing drones based on size or weight, the proposal takes a risk-centered approach, focusing on the risk of actual operations.[56] This means that even a smaller drone can be the subject of regional regulation, where such a need arises. Where higher-risk drone operations are undertaken, certification will be required according to the proposal.[57]

Surveillance and privacy issues are generally outside the scope for civil aviation authorities, as their competence is purely within air safety.[58] However, the proposed Regulation on Civil Aviation deals explicitly with these aspects; while no update of the privacy framework is deemed necessary, the proposal does connect the areas together somewhat, for example, by suggesting closer collaboration between national aviation authorities and national data protection authorities.[59]

The approach that has been adopted by the EU is one clearly updated from that of the previous Regulation on Civil Aviation of a decade ago, due in no small part to developments in technology.[60] While it is a step in the right direction, attempting to update aviation rules to include drones to a much larger scale, some aspects are still outside the scope of civil aviation authorities, and will need to be dealt with through other means. For example, liability resulting from harm to people or property caused by drones may not be included within aviation regulations, even those that have been updated for unmanned aircraft. Existing provisions of tort law may or may not include liability through remotely operated vehicles; such provisions could be updated to consider not only drones but also driverless cars and other vehicles.[61] Reckless operation of a drone, which can be compared to dangerous driving, would also need to be regulated in a similar way. Transport offences may require

---

[51] European Aviation Safety Agency (2015).

[52] See European Aviation Safety Agency (2017).

[53] European Commission (2015b).

[54] European Commission (2015b), p. 2 and European Commission (2015a). Annex IX of the proposal deals specifically with drones, entitled Essential requirements for unmanned aircraft.

[55] Article 45 proposed Regulation on Civil Aviation.

[56] Article 46 proposed Regulation on Civil Aviation.

[57] See, further, European Council, Drones: reform of EU aviation safety, http://www.consilium.europa.eu/en/policies/drones/. Accessed 4 February 2018.

[58] See, further, Clarke (2014a), p. 296.

[59] European Commission (2015b), p. 7.

[60] The proposed Regulation on Civil Aviation also refers to aspects such as the environment, which will be discussed in Sect. 4.1 below.

[61] This is an area that can be complex to regulate from an EU perspective, as insurance and liability regimes differ between EU Member States, see further Rule (2015), pp. 196–197.

amendments to include, for example, an offence of "dangerous drone operation" or "dangerous unmanned aerial vehicle operation."

All of these aspects are traditionally dealt with at a national level and therefore pose a challenge for the EU similar to that it is currently trying to address in relation to aviation regulations, namely avoiding a fragmented regulatory approach.

3.4 Additional Regulatory Issues

The regulation of drones is no simple task, and as with many developing technologies leads to questions relating to a number of fields of law. The resulting complexity of regulation is without doubt a challenge, and although a number of steps have already been taken to update the legal framework and encourage drone innovation, potential loopholes remain. Although this chapter has focused on surveillance, privacy and aerial regulations, other areas are also affected by the growing use of drones and may need regulatory review. These areas include product liability, property law, and corporate espionage.[62] For example, property law may need adjustment to cover situations where drones fly too close to property, in the form of aerial trespass provisions.[63] Where drones malfunction or operator error results in harm to persons or property, insurance and liability laws come into play, as mentioned above.[64] It will be interesting to see if and how these issues are dealt with from a regulatory standpoint in the future.

4 The Future for Drones

Although drones have not yet been adopted in all sectors and we have not yet reached the stage of public drone airways or toll systems for drones,[65] there is a clear need to assess the current regulatory environment. Deployment of drone technology is becoming more widespread and questions are being raised as to how drones can be integrated safely into society.

---

[62]Much of the regulation within these fields is based on EU framework legislation and should therefore be possible at a national level.

[63]For discussion of property law and aerial trespass in a US context, see Rule (2015), pp. 186–188.

[64]For a discussion of EU liability laws in relation to automated cars, see Schellekens (2015), pp. 513–517.

[65]See Rule (2015), pp. 196–197.

4.1 Potential Risks and Hidden Benefits

While there clearly are potential risks involved in drone usage, these risks can be mitigated by making use of available technology for good, for example, by improving privacy and safety standards. Building safeguards into drones, programming them to do such things as avoid no-fly zones and automatically blur privacy-sensitive information, goes some way to minimizing the risks of surveillance technology. The benefits of drone usage are, however, much more widespread, impacting the environment, society as a whole and the economy.

Drone deliveries have the ability to reduce the amount of pollution traditional deliveries entail, leading to a clear beneficial environmental impact. This is particularly the case as the technology advances, allowing drones to fly further, or being powered by renewable energy sources.[66] Danger to human life can be reduced through the use of drones in dangerous situations, such as after natural disasters and in remote areas. From an economic standpoint, drones have the potential to reduce the costs of, for example, transportation or human resources. Developments in drone technology can therefore be a force for good, much like the fictional robot *WALL-E.*

4.2 Flying High in the EU?

The EU is, in general, well placed for coherent drone regulation. The current and developing framework within the areas of surveillance, privacy and aviation is capable of providing a balance between regulation and the encouragement of innovation. Rather than developing a specific regulatory solution for drones, the EU approach is to integrate drones into the existing legislative framework; this integration is based on what the drone does and its behavior, rather than on the technology itself. This approach provides the necessary flexibility for the framework to adapt to new risks and benefits of drone technology.

The proposed Regulation on Civil Aviation, besides updating somewhat outdated legislation, should result in the encouragement of drone usage both within Member States and between EU countries. Care must still be taken to ensure that Member States' domestic rules do not differ too greatly or conflict with each other so as to stifle innovation and drone usage between countries. In addition, where regulatory questions fall to the national jurisdictions, such as in the area of surveillance, individual countries must make sure to achieve the correct balance between legitimate interests and the protection of information. It is encouraging, however, that the EU is already addressing the issue of drones to such a great extent; as a result, Member States do not need to find their own specific and tailored solutions to drone usage, a situation that would most likely lead to confusion on the EU market place. The proposed Regulation on Civil Aviation, together with the new privacy framework

[66] Solar powered drones have been envisaged for a number of years, see, e.g., Boeing (2010).

provided by the GDPR, may not address all concerns, but certainly go some way towards drones flying high within the EU.

References

Article 29 Data Protection Working Party. (2015). Opinion 01/2015 on privacy and data protection issues relating to the utilisation of drones.

BBC News. (2016). Dubai airport grounds flights due to 'drone activity', 28 September 2016, http://www.bbc.com/news/world-middle-east-37493404. Accessed 4 Feb 2018.

Boeing. (2010). Boeing wins DARPA vulture H program, 16 September 2010, http://boeing.mediaroom.com/index.php?s=20295&item=1425. Accessed 4 Feb 2018.

Clarke, R. (2014a). Understanding the drone epidemic. *Computer Law & Security Review, 30*(3), 230–246.

Clarke, R. (2014b). The regulation of civilian drones' impacts on behavioural privacy. *Computer Law & Security Review, 30*(3), 286–305.

Corcoran, M. (2014). Chinese manufacturer programs Phantom drones with no-fly zones to protect Australian Airports, ABC News, 14 April 2014, http://www.abc.net.au/news/2014-04-14/chinese-made-drones-programmed-with-no-fly-zones/5388356. Accessed 4 Feb 2018.

Cracknell, A. P. (2017). UAVs: Regulations and law enforcement. *International Journal of Remote Sensing, 38*(8–10), 3054–3067.

Deutsche Post DHL Group. (2016). Successful trial integration of DHL Parcelcopter into logistics chain, Press Release, 5 September 2016, http://www.dpdhl.com/en/media_relations/press_releases/2016/successful_trial_integration_dhl_parcelcopter_logistics_chain.html. Accessed 4 Feb 2018.

Dorling, K., et al. (2017). Vehicle routing problems for drone delivery. *IEEE Transactions on Systems, Man, and Cybernetics: Systems, 47*(1), 70–85.

European Aviation Safety Agency. (2015). Introduction of a regulatory framework for the operation of unmanned aircraft, technical opinion, RMT.0230, 18 December 2015), https://www.easa.europa.eu/system/files/dfu/Introduction%20of%20a%20regulatory%20framework%20for%20the%20operation%20of%20unmanned%20aircraft.pdf. Accessed 4 Feb 2018.

European Aviation Safety Agency. (2017). EASA publishes a proposal to operate small drones in Europe, Press Release, 5 May 2017, https://www.easa.europa.eu/newsroom-and-events/press-releases/easa-publishes-proposal-operate-small-drones-europe. Accessed 4 Feb 2018.

European Commission. (2014). A new era for aviation: Opening the aviation market to the civil use of remotely piloted aircraft systems in a safe and sustainable manner, Communication from the Commission to the European Parliament and the Council, COM (2014) 207 final.

European Commission. (2015a). Annexes to the proposal for a regulation of the European Parliament and of the Council on Common Rules in the Field of Civil Aviation and Establishing a European Union Aviation Safety Agency, and Repealing Regulation (EC) No. 216/2008 of the European Parliament and of the Council, COM (2015) 613 final.

European Commission. (2015b). Proposal for a regulation of the European Parliament and of the Council on Common Rules in the Field of Civil Aviation and Establishing a European Union Aviation Safety Agency, and Repealing Regulation (EC) No. 216/2008 of the European Parliament and of the Council, Explanatory Memorandum, 2015/0277 (COD).

Home Office. (2013). Surveillance camera code of practice, June 2013, https://www.gov.uk/government/publications/surveillance-camera-code-of-practice. Accessed 4 Feb 2018.

Luppicini, R., & So, A. (2016). A technoethical review of commercial drone use in the context of governance, ethics, and privacy. *Technology in Society, 46*, 109–119.

McNeal, G. S. (2016). Drones and the future of aerial surveillance. *George Washington Law Review, 84*(2), 354–416.

Rule, T. (2015). Airspace in an age of drones. *Boston University Law Review, 95,* 155–208.
Sandbrook, C. (2015). The social implications of using drones for biodiversity conservation. *Ambio—A Journal of the Human Environment, Springer, Netherlands, 44*(Suppl. 4), 636–647.
Schellekens, M. (2015). Self-driving cars and the chilling effect of liability law. *Computer Law & Security Review, 31*(4), 506–517.
Shelley, A. V. (2016). Application of New Zealand Privacy Law to drones. *Policy Quarterly, Victoria University of Wellington, 12*(2), 73–79.
SVT News. (2017). Drönare som stoppar flygtrafiken – ett växande problem, 31 May 2017, www.svt.se/nyheter/lokalt/stockholm/dronare-som-stoppar-flygen-ett-vaxande-problem. Accessed 4 Feb 2018.
The Guardian. (2016). First passenger drone makes its debut at CES, 7 January 2016, https://www.theguardian.com/technology/2016/jan/07/first-passenger-drone-makes-world-debut. Accessed 4 Feb 2018.
The Independent. (2017). Drone at Gatwick Airport closes runway and causes flights to be diverted, 2 July 2017, http://www.independent.co.uk/news/uk/home-news/gatwick-airport-drone-latest-disruption-flight-diverted-easyjet-runway-closed-a7819841.html. Accessed 4 Feb 2018.
The Verge. (2016). GoPro is recalling its Karma drone, 8 November 2016, https://www.theverge.com/2016/11/8/13569730/gopro-karma-drone-recall-announced. Accessed 4 Feb 2018.
The White House. (2015). Promoting economic competitiveness while safeguarding privacy, Civil Rights, and Civil Liberties in Domestic Use of Unmanned Aircraft Systems, Presidential Memorandum, 15 February 2015, https://obamawhitehouse.archives.gov/the-press-office/2015/02/15/presidential-memorandum-promoting-economic-competitiveness-while-safegua. Accessed 4 Feb 2018.
Volovelsky, U. (2014). Civilian uses of unmanned aerial vehicles and the threat to the right to privacy—An Israeli case study. *Computer Law & Security Review, 30*(3), 306–320.
Weismann, A. L. (2015). How do the FBI's self-described domestic law enforcement drones implicate foreign intelligence? CREW: Citizens for responsibility and ethics in Washington, 6 January 2015, https://www.citizensforethics.org/just-how-does-the-fbis-drone-program-a-self-described-domestic-law-enforcem/. Accessed 4 Feb 2018.
West, J. P., & Bowman, J. S. (2016). Domestic use of drones: An ethical analysis of surveillance issues. *Public Administration Review, The American Society for Public Administration, 76*(4), 649–659.

Profiling and Automated Decision-Making: Legal Implications and Shortcomings

Stefanie Hänold

Abstract The increased use of profiling and automated decision-making systems raises a number of challenges and concerns. The underlying algorithms embody a considerable potential for discrimination and unfair treatment. Furthermore, individuals are treated as passive objects of algorithmic evaluation and decision tools and are unable to present their values and positions. They are no longer perceived as individuals in their own right: all that matters is the group they are assigned to. Profiling and automated decision-making techniques also depend on the processing of personal data, and a significant number of the available applications are highly privacy-intrusive. This article analyses how the European General Data Protection Regulation (GDPR) responds to these challenges. In particular, Art. 22 GDPR, which provides the right not to be subject to automated individual decision-making, as well as the information obligations under Art. 13 (2) (f) and Art. 14 (2) (g) GDPR and the access right under Art. 15 (1) (h) GDPR, will be examined in detail. General data protection principles, particularly the principle of fairness, as well as specific German scoring provisions and anti-discrimination rules, are looked at, too. In conclusion, various shortcomings of the present legal framework are identified and discussed and a short outlook for potential future steps presented.

Keywords Profiling · Automated decision-making · Algorithm · Explanation General Data Protection Regulation (GDPR)

S. Hänold (✉)
Institute for Legal Informatics, Leibniz Universität Hannover, Hannover, Germany
e-mail: haenold@iri.uni-hannover.de

M. Corrales et al. (eds.), *Robotics, AI and the Future of Law*, Perspectives in Law, Business and Innovation, https://doi.org/10.1007/978-981-13-2874-9_6

1 Introduction

Big Data[1] analytics now create enormous economic revenues, and it is expected that these will even grow by more than 50% over the next five years.[2] Besides economic wealth, societal challenges may be tackled with Big Data instruments, e.g., Smart Grids are now widely used to reduce energy consumption, and it is envisaged that Big Data technologies will be used to manage the pollution of the environment in the future.[3] Big Data development is triggered, on the one hand, by the enormous growth of data pools, fed among other things from day to everyday actions of the population, e.g., using mobile phones, online banking, Internet shopping or ostensibly free services, such as social networks. Additionally, the Industry 4.0 development[4] contributes to the massive increase of data. On the other hand, the technological means to process huge amounts of data sets in real time and to extract new knowledge from these data have caused a societal transition which is in its future extent unknown.

However, certain forms of usage of Big Data technologies are already heavily discussed. One area of discussion centres around the use of profiling techniques and the use of automated decision-making. Individuals are increasingly confronted with decisions based on automated processing. Some of these decisions can be vital to the individuals concerned, e.g., when it comes to decisions concerning creditworthiness, housing, employment or being a crime suspect.[5] At first one might assume that the automation of processes increases objectivity and fairness in decision-making as the personal sensitivities of natural persons no longer have a bearing and therefore automated decisions may be of advantage to the concerned individuals. Nonetheless, taking a closer look, it is evident that the underlying algorithms of automated decision-making programs have immense discriminatory potential. Furthermore, the increased collection of personal data of individuals for profiling purposes also endangers the individual's right to data protection and privacy. It also engages the right to personality if individuals are ever more exposed to decisions based on automated processing. This chapter will take a closer look at the phenomenon of profiling and automated decision-making. To begin, a short explanation will be given as to what profiling and automated decision-making is, including some illustrative examples to provide a more concrete picture. In the process, the role of algorithms will also be outlined. After focusing on the issues caused by increased use of profiling and automated decision-making systems for the concerned individuals and society, the legal regulation of profiling and automated decision-making will be analysed. Finally, it will be discussed whether the current legal norms provide an adequate response to the outlined issues.

---

[1] For a clarification of the term Big Data, see Forgó et al. (2017), pp. 20–22.

[2] Hacker and Petkovka (2017), p. 4.

[3] See Footnote 2.

[4] Hofmann (2016) gives a brief explanation of the characteristics of the Industry 4.0 development, pp. 12–13.

[5] Edwards and Veale (2017), p. 19 and Vedder and Naudts (2017), p. 207.

2 What Is Profiling and Automated Decision-Making?

2.1 Profiling, Automated Decision-Making and Algorithms

This chapter aims to provide an overview of profiling and automated decision-making. As algorithms are necessary components for profiling and automated decision-making and a lot of the issues created are traceable to their algorithmic nature, more light will be shed on what algorithms are and why they are key components in Big Data development.

2.1.1 Profiling and Automated Decision-Making

Profiling is characterized by the fact that service providers or other data controllers process personal data automatically with the help of algorithms to evaluate certain personal aspects relating to an individual, in particular in order to analyse or predict the conduct of a person, e.g., concerning his or her performance at work, economic situation, health, personal preferences, interests, reliability, behavior, location or movements.[6] At the same time, profiles are an interpretation of data relating to a specific individual and thus include an evaluative element.[7] Often data from various sources are used and inferences are made based on qualities of other people that seem statistically alike (the probability based on a group).[8] One example for profiling is credit scoring where mathematical-statistical procedures are used to determine or predict the economic situation and reliability of a person.[9] Depending on the purpose of the profiling, it can serve also the interests of the community and those of the concerned individual.[10] For example, credit scoring by the German credit investigation company SCHUFA protects, on the one hand, credit businesses from losses in the loan business. On the other hand, the aim is also to protect consumers against over-indebtedness.[11]

Automated decisions are decisions made by technological means. Often automated decision-making systems include the construction and evaluation of profiles.[12] Then, the automated decision depends on the result of the profiling.[13] However, automated decisions can also take place without profiling and profiling techniques can be

[6] Art. 4 (4) GDPR; Martini (2018), Art. 22 margin note 21.

[7] Art. 29 Data Protection Working Party (2018), p. 7 and Martini (2018), Art. 22 margin note 21.

[8] Art. 29 Data Protection Working Party (2018), p. 7.

[9] Buchner (2018), Art. 22 margin note 22.

[10] Hladjk (2017), Art. 22 margin note 4.

[11] Kamlah (2003), p. V.

[12] Steppe (2017), p. 783.

[13] Buchner (2018), Art. 22 margin note 4.

used without making automated decisions.[14] In such cases profiles may, for example, serve as a decision support for a human taking the final decision.[15]

2.1.2 Algorithms

Algorithms have become central steering tools of the digitalized society and influence more and more people's lives, e.g., they are key components for profiling and automated decision-making systems.[16] Algorithms as such can in general be described as step-by-step instructions for the solution of a mathematical problem.[17] They are no phenomenon of the digital age, as the first mathematical algorithms date back about 2000 years.[18] However, only with the use of algorithms it is possible to find meaning in the massive data sets of today.[19] In order to make sense of these huge amount of data, algorithms search for patterns, correlations and commonalities in the data sets in real time.[20] Decision-making algorithms regularly base decisions on correlations—relationships between variables—discovered by algorithms used in Big Data analytics.[21] For example, if it is statistically proven that on a certain date more people want to travel to a certain place, the algorithm of a travel company may charge higher prices for that travel destination on that specific date.[22] The possibilities for employing algorithms increase and change every day. In the last couple of years, the move from manually specified rule based algorithms towards complex machine learning algorithms has enabled the modelling of complex social phenomena with much greater accuracy and therefore a higher level of usability.[23] These days neural networks are capable of generating correct results just on what they learned from training data.[24] Programmers also work on neuronal networks which are able to write their own programs independent from the programming by their human developers.[25] Looking at the recent steps in developing and application of AI, a lot of further development is to be expected.[26]

---

[14] Art. 29 Data Protection Working Party (2018), p. 8 and Kamlah (2016), Art. 22 margin note 2.

[15] Art. 29 Data Protection Working Party (2018), p. 8, Edwards and Veale (2017), p. 19 and Kamlah (2016), Art. 22 margin note 2.

[16] McLellan (2016).

[17] Martini (2017), p. 1017.

[18] Martini (2017), p. 1017 and Ernst (2017), p. 1026.

[19] Vedder and Naudts (2017), p. 206.

[20] Vedder and Naudts (2017), p. 210.

[21] Ernst (2017), p. 1028.

[22] See Footnote 21.

[23] Edwards and Veale (2017), p. 25.

[24] McLellan (2016) and Hoffmann-Riem (2017), p. 3.

[25] Datatilsynet, The Norwegian Data Protection Authority (2018), p. 6.

[26] Hoffmann-Riem (2017), p. 3 and McLellan (2016).

2.2 *Examples of Automated Decision-Making and Profiling Systems*

As applied to individual persons, profiling and automated decision-making systems are omnipresent and serve a broad variety of purposes. Increasingly, they affect everybody's lives both in a private and public context.[27] In the following, some examples of automated decision-making, including profiling, are introduced.[28]

2.2.1 Real-Time-Loans via a Smartphone-App

One example of automated decision-making, including profiling, are loan approvals via smartphone apps. The credit assessment runs fully automatically and the customer receives an offer within minutes. At the beginning, customers indicate the desired amount of money and answer a question catalogue regarding their personal financial situation. Additionally, the financial history and information from credit agencies are automatically evaluated. Depending on their score, they receive an offer or the application is denied.[29]

2.2.2 Insurance Premiums

Car insurance companies face very tough market conditions as the market is saturated and is, therefore, characterized by heavy price and predatory competition. Therefore, insurance companies rely, as much as possible, on a differentiated premium model. In order to be able to offer such a model, they need, in turn, to know the individual risk of insurance holders.[30] For this, in turn, they need to have a clear picture of their customers driving behavior. For that reason, insurance companies have started to offer insurance contracts where they collect driver's data in respect of speed, style, day and night driving with telematics devices built-in the car.[31] The collected data is analyzed and premiums are based on the driver's style.[32] Moreover, such data provides the only basis for a decision about the premium.[33]

---

[27] Edwards and Veale (2017), p. 19.

[28] Whether these examples constitute automated decision-making in the sense of Art. 22 (1) or (4) GDPR is discussed in Sect. 4.1.1.1.

[29] IT Finanzmagazin (2017).

[30] Schwichtenberg (2015), p. 378.

[31] Lüdemann et al. (2014), p. 304 and Schwichtenberg (2015), p. 379.

[32] Schwichtenberg (2015), p. 378; Lüdemann et al. (2014), pp. 302–303.

[33] Lüdemann et al. (2014), p. 304.

2.2.3 Recruiting

In the field of job applications, profiling and automated decision-making systems are also used, even if, for now, the final decision about whom to employ remains with a natural person. For example, applicants can be judged by software which evaluates their personality by language analysis[34] or psychographs are produced by assessing profiles from profession-oriented networks. This way it can be figured out in a cost-saving manner whether the applicant is suitable for the position, what development potential an applicant has, and if he or she matches the corporate culture.[35] The individuals responsible for the final choice will consider the results of such an analysis, but will also make their judgement on the basis of their own experience and personal impression of the candidate. In contrast to this, CV filtering systems can be used by companies, especially if they receive a large number of applications. In these cases, applicants who do not achieve a certain score will be automatically denied progression to the next round of the selection process.

2.2.4 Personalized Pricing

Online vendors are able to categorize by means of the data they collect from their customers whether a customer is to be classified as an "affluent customer" or a "budget conscious customer" and can quickly adapt prices accordingly.[36] For personalized pricing, a company can use a broad range of information which can be provided voluntarily, like customer account data, but information can also be provided unintentionally or involuntarily, e.g., the IP address which gives information about the country and region where the customer lives and the Internet provider.[37] It is also possible to find out which computer type the customer uses and what kind of Internet connection they have.[38] Browser cookies may also give online vendors the possibility to infer the probability of a decision to buy a product at a certain price point.[39] Third parties, e.g., advertising networks, could also provide valuable data for price discrimination, for example, information that represents the customer's online behavior which can be gathered by using cookies.[40]

---

[34] Der Tagesspiegel (2018) and Ernst (2017), p. 1026.

[35] Der Tagesspiegel (2018) and Schönhaar (2018).

[36] Steppe (2017), p. 781.

[37] Borgesius Zuiderveen and Poort (2017), p. 350.

[38] See Footnote 37.

[39] James (2015).

[40] See Footnote 37.

3 Societal Challenges of Profiling and Automated Decision-Making

3.1 Discrimination

It is often suggested that the use of profiling and automated decisions may be the answer to social grievances caused by prejudice and a lack of openness. For example, in application procedures, recruiters often make decisions based on stereotypes and sometimes they may be unaware that they are doing so. One of the possible consequences of such discrimination, for example, is that people with a foreign-sounding name, although having the same expertise as other applicants, are not invited to a job interview.[41] It is argued that the use of e-recruiting tools can make the recruitment process more effective, objective and less vulnerable to such cognitive biases.[42]

However, a closer look reveals that the "algorithm solution" needs to be treated with caution. Since algorithms are designed and implemented by humans, their design and use are inevitably influenced by the personal attitudes, values, inclinations, and tendencies of those persons that program and use the algorithms for a specific purpose.[43] Decisions made by algorithms may seem objective because the decision is based on a huge data basis. However, humans decide which criteria are used for the decision-making and what weight is assigned to them.[44] In case an algorithm is trained on a specific data basis, inherent biases are likely to be re-entrenched.[45] For instance, if a CV filtering system is developed based on the success rates of former applicants, presumably the biases of the persons that have filtered former CVs implicitly will be taken over into the automated filter system, and this may happen unintentionally.[46] This shows how problematic it can be to "train" algorithms with past data as discrimination from previous decision-making processes are likely to be repeated or even aggravated.[47] Plenty of other examples for discriminating algorithms have been discussed, e.g., the software Compas which has been used in the U.S. to predict the likelihood of criminals reoffending. For black criminals, the predictions deviated—in comparison to white criminals—significantly from the actual recurrence rate.[48] Building non-discriminatory systems has proven a formidable challenge because it is not sufficient to exclude obvious discriminatory values, e.g., race or gender. Developers also need to avoid indirect discrimination by excluding values that may function as proxies for the omitted and discrimina-

[41] Kramer (2018).

[42] Kramer (2018) and Ernst (2017), pp. 1027–1028.

[43] Ernst (2017), p. 1029 and Vedder and Naudts (2017), p. 209.

[44] Ernst (2017), p. 1029.

[45] McLellan (2016), Edwards and Veale (2017), p. 28 and Schermer (2011), p. 47.

[46] Edwards and Veale (2017), p. 28.

[47] Edwards and Veale (2017), p. 28 and Ernst (2017), pp. 1028–1029.

[48] Martini (2017), p. 1018.

tory characteristics.[49] Examples are distance from home to work or criminal records which can both correlate with racial background, or individual working and holidays which allow inferences about religious beliefs.[50]

3.2 Objectification and Opacity

One of the characteristics of profiling algorithms is that the individual whose profile is created is evaluated by the probability of a group. Since algorithms only use correlations and statistical models, the persons concerned are only perceived as members of a group and not as individuals in their own right.[51] This is a matter of concern because group profiles may be valid for the group and individuals as members of the group, but not for the individual as such.[52] In addition, individuals are treated as the passive object of algorithmic evaluation and decision tools and are not given the opportunity to present their own distinct values and positions.[53] Objectification is enhanced by the fact that profiling and automated decision-making are—for the concerned individuals—a non-transparent process in which the profiling and decision steps remain hidden and the individual concerned is only presented with the result.[54] While decisions of rule-based algorithms are, in principle, explainable[55], controllers that use the algorithm have little interest in providing such information about the logic behind the decision as they fear that individuals could "game" the system or they might lose their competitive advantage.[56] However, even if systems are, in principle, explainable, the underlying processes are often of a highly complex and opaque nature.[57] The possibilities of insight into sociological and psychological relations which today's algorithms possess go far beyond human perception.[58] Algorithms recognise widely varied criteria as the basis of a decision and the human capability of understanding the algorithm's decision process is limited.[59] When it comes to neural networks[60] and other forms of machine learning, the additional

[49]Edwards and Veale (2017), p. 29 and Ernst (2017), p. 1032.

[50]Hacker and Petkovka (2017), p. 8.

[51]See Footnote 21.

[52]Schermer (2011), p. 47, provides an explanatory example.

[53]Ernst (2017), p. 1030 and Jandt (2015), p. 8.

[54]See also Schaar (2016), Edwards and Veale (2017), pp. 19–20 and Hladjk (2017), Art. 22 Rn. 3.

[55]A decision tree is a simple model which provides a high degree of transparency. For further information, see Datatilsynet, The Norwegian Data Protection Authority (2018), p. 13. An illustrative example is provided by Binns (2017).

[56]See Footnote 20.

[57]Vedder and Naudts (2017), pp. 208, 217, Hoffmann-Riem (2017), p. 29 and Martini (2017), p. 1018.

[58]Ernst (2017), pp. 1028–1029, Vedder and Naudts (2017), p. 210 and Clifford Chance (2017).

[59]Ernst (2017), p. 1030 and Clifford Chance (2017).

[60]An explanation of the term "neural networks" can be found in Datatilsynet, The Norwegian Data Protection Authority (2018), p. 14.

problem occurs that the internal decision-making processes are not transparent even to their developers who are not able to tell exactly how a certain decision has been generated.[61] "Explainable AI" is a hot topic of inquiry as awareness arises that the use of self-learning and self-optimizing systems in certain contexts, e.g., the finance or medical sector, cannot be pursued if the reasons for the decisions of the algorithms can no longer be verified.[62] "Explainable AI" would not just be of benefit for the concerned individuals. Furthermore, system designers could also make use of them. Such systems—because of their "probabilistic nature"—are expected to fail in some instances[63] and with the help of explanations, the "up-to-datedness" and reliability of a system could thus be verified.

3.3 *Privacy and Autonomy*

Big Data analytics make peoples personalities and lives extremely transparent. Especially—but not exclusively—business corporations tend to collect as much personal data as they can in order to analyze them and use the results to increase their profits. Such data is often used without consideration of the context the data was previously collected and processed for and the data is often also sold to other companies for other purposes.[64] Under such conditions, the majority of people now feel like they have lost control over their personal information.[65] Many people lack the knowledge of how to protect their personal data[66] or a willingness to investigate the consequences of leaving personal traces with providers.[67] Habituation to use certain services and so-called "lock-in" effects of social networks also make it easy for "data leeches" to collect even more information about their users.[68] As a result of profiling, knowledge or "inferences" about the group, the individual is assigned to, is integrated to hypothesize about that individual's likely attributes or behavior, which might be, for example, products and services likely to be of interest to them or the social connections they wish to develop or medical conditions they may suffer from in the future.[69] In that sense, the process of profiling creates new personal data and the person concerned may not be aware of that generated data.[70] Therefore, the profile data

[61] McLellan (2016), Knight (2017), Ernst (2017), p. 1027, Martini (2017), pp. 1018–1019 and Datatilsynet, The Norwegian Data Protection Authority (2018), p. 12.

[62] Stolberg and Ceccotti (2018), Knoche (2018) and Holzinger et al. (2017).

[63] Edwards and Veale (2017), p. 54.

[64] Edwards and Veale (2017), pp. 32–33 and Hoffmann-Riem (2017), p. 23.

[65] Edwards and Veale (2017), p. 33 and Hacker and Petkovka (2017), p. 7.

[66] Hacker and Petkovka (2017), p. 7.

[67] Hildebrandt (2009), p. 243.

[68] See Footnote 66.

[69] Edwards and Veale (2017), p. 32.

[70] Art. 29 Data Protection Working Party (2018), p. 9.

constitute an "invisible visibility" for the individuals that are subject to profiling.[71] The data is of high sensitivity as the future conduct of the concerned individuals can be more accurately predicted and knowing people's needs and preferences makes it easier to influence or steer them. This development should be viewed with concern as profiling techniques make it possible to trace persons to an extent which would have been unthinkable in the past and the people most affected are, for the most part, unaware of the manipulative processes to which they are now routinely exposed.[72]

4 Legal Regulation

The following section presents how profiling and automated decision-making is captured from the perspective of the European Data Protection law as well as specific German provisions with regard to scoring and German anti-discrimination rules.

4.1 The European General Data Protection Regulation

In recent years, the European data protection regime has undergone major reform. As a consequence of this, the new GDPR applies from 25 May 2018 replacing the Data Protection Directive from 1995. Due to its character as a European regulation, the provisions of the GDPR are directly applicable in all the European Member States; though the Regulation also provides numerous enabling clauses that permit national deviations.[73] The GDPR, which principally provides statutory requirements for the processing of personal data with automated means in an European context,[74] contains a number of provisions that have to be considered with respect to automated decision-making and profiling that involve the processing of personal data. One of the key provisions in this regard is Art. 22 GDPR which determines under which conditions the use of automated decision-making is permitted. However, as regards the admissibility of processing personal data for the purpose of automated decision-making, the regular provisions of the Regulation must also be complied with.[75] For example, the processing of the personal data must be based on a legal ground which follows from Art. 5 (1) (a) and Art. 6 or Art. 9 GDPR. Art. 22 GDPR itself does not constitute a legal ground for the processing of personal data as such.[76] This also applies to profiling processes that involve the processing of personal data whether an automated decision follows the profiling process or not. In addition to Art. 22 GDPR,

[71] Hildebrandt (2009), p. 242.

[72] Hildebrandt (2009), p. 244.

[73] Kühling and Martini (2016), pp. 448–449.

[74] For details regarding the material and territorial scope see Art. 3 and 4 GDPR.

[75] Kamlah (2016), Art. 22 margin note 2; Schulz (2017), Art. 22 margin note 4.

[76] Schulz (2017), Art. 22 margin note 4.

especially the information obligations towards the data subject, regulated in Art. 13 and 14 GDPR, as well as the data subject's access right, as provided for under Art. 15 GDPR, are of importance.

4.1.1 The Right not to Be Subject to a Decision Solely Based on Automated Processing

According to Art. 22 (1) GDPR, the data subject shall have the right not to be subject to a decision based solely on automated processing, including profiling, which has legal effects for him or her, or similarly significantly affects him or her. Although Art. 22 (1) GDPR is formulated as a right of the data subject, it is to be interpreted as an objective prohibition and does not depend on its exercise by the concerned data subject.[77] The prohibition is not absolute, however. In Art. 22 (2) GDPR the European legislator has provided some exceptions when automated decision-making is still admissible. Since these exemptions are fairly comprehensive, it could be argued that the relationship between the rule and the exception are, in fact, reversed.

4.1.1.1 Scope of Application

4.1.1.1.1 A Decision Is Based Solely on Automated Processing

Looking first at the scope of application of Art. 22 (1) GDPR, this requires, inter alia, that the decision is based solely on automated processing. In the case that no human being is involved in the automated decision-making process, this condition is easily fulfilled. Recalling the example in Sect. 2.2.1 of the real-time-loan granting via a smartphone-app, the applicant's creditworthiness is evaluated exclusively by the bank's software and the decision to offer a credit and under which conditions is based solely on the results of the previous automated assessment. No human intervention is involved. With respect of the further example of the setting of the car insurance fee, an automated decision is also made solely on the basis of the score and therefore this decision also qualifies easily as a decision based solely on automated processing. In contrast, considering the recruiting example, from Sect. 2.2.3, the software only evaluates the applicant's personality by language analysis and creates a profile with respect to how well the applicant might fit into the corporate culture. The results of the analysis are in the end not decisive and only one factor that the recruitment staff use in making the final decision as to which applicant will be offered the job. Here, unlike the first two illustrated examples, the decision is not based solely on automated processing. Nevertheless, in a case where a human person makes the final decision, but routinely simply "rubber stamps" the result of the profiling process,

[77] Art. 29 Data Protection Working Party (2018), p. 19, Schulz (2017), Art. 22 margin note 5; Martini (2018), Art. 22 margin note 1. A different view is expressed by Kamlah (2016), Art. 22 margin note 4. Wachter et al. (2017), p. 95, conclude that the formulation chosen is critically ambiguous.

this would still qualify as a decision based solely on automated means.[78] For one thing, the wording refers to decisions based solely on automated processing and not "solely automated decision-making."[79] Moreover, it would otherwise be too easy for data controllers to circumvent the prohibition to use automated decision-making, simply by involving staff as "decision-makers" who automatically approve the result of the fully automated profiling process.[80] A different evaluation is only justified when the human decision-maker has the authority, competence and time to change the decision considering all the relevant data.[81] Consequently, in cases where pre-established guidelines leave no room for own judgements, the involvement of a human person would not hinder Art. 22 (1) GDPR from applying.[82] The same applies if human capacity to retrace the crucial reasons for the decision has reached its limits as, for example, in a case where the underlying algorithmic processes and data are too complex to comprehend. The issue of "automation bias" needs to be considered in that context as well. Automation bias is a psychological phenomenon and may lead to the over-reliance on decision support systems by humans.[83] The more complex the factors are that play a role for the decision, the more humans may be inclined to trust the automated system rather than their own judgement.[84]

4.1.1.1.2 Legal Effect or Similarly Significant Effect

Art. 22 (1) GDPR also states, that to fall under special regulation, the decision solely made by automated means should have a legal effect for the concerned person or similarly significantly affect him or her. A legal effect is produced by the decision when a legal position is established, changed or cancelled or if the decision infringes on the right of a person.[85] As an example, the cancellation of an existing contract could be mentioned. In case a legally binding offer is made to the individual, this also has a legal effect, as the individual is entitled to accept that offer with the consequence of a legally binding contract.[86] However, in shops and online shops the vendors usually only make an invitation to treat which has no legal effect for the customer. In these cases, the customer makes a legally binding offer and if the vendor accepts, both enter into a legally binding contractual relationship. Where there is only an invitation to treat, which has as such no legal effect, it must be

---

[78] Helfrich (2017), Art. 22 margin note 44 and Malgieri and Comandé (2017), p. 9. A more restrictive view is taken by Wachter et al. (2017), p. 92.

[79] Malgieri and Comandé (2017), p. 251.

[80] Lüdemann et al. (2014), p. 304 and Steppe (2017), p. 783.

[81] Art. 29 Data Protection Working Party (2018), p. 21 and Schulz (2017), Art. 22 margin notes 14–15.

[82] See Footnote 79.

[83] Edwards and Veale (2017), p. 45.

[84] See also Vedder and Naudts (2017), pp. 216–217.

[85] Buchner (2018), Art. 22 margin note 24.

[86] Steppe (2017), p. 784. A more restrictive view is taken by von Lewinski (2018), Art. 22 margin note 28.

further investigated whether the decision similarly significantly affects the concerned individual. This criterion is open to interpretation and it is questionable how far subjective perceptions of the concerned individuals are to be considered or if the assessment rather depends on objective criteria. In Recital 71 of the GDPR, the European legislator mentions the rejection of an application for a credit and also job denials in e-recruitment procedures, which is plausible as such decisions have considerable economic significance for the concerned individuals.[87] The Art. 29 Working Party[88] mentioned various contexts for decisions where automated decisions may be considered sufficiently significant to meet the threshold, e.g., when it comes to access to health services or education.[89]

It is questionable how price discrimination[90] is to be evaluated in this light, as online offers are generally regarded as invitations to treat and produce no legal effect. Therefore, it must be verified whether these can similarly significantly affect individuals.[91] In general, the principle of the freedom of contract applies and there is no obligation on the side of the vendors to sell their products to each customer for the same price.[92] Admittedly, customers may feel that some injustice has occurred. However, the impact on the individual must arguably exceed this, so as to amount to a considerable impairment.[93] It appears to make sense to consider the individual circumstances of each case, e.g., in the case of price discrimination the price difference between the cheapest offer and the offer in question.[94] In the case of prohibitively high prices that force customers to refrain from a purchase, the price decision may have significant effect on the individual.[95] Certain forms of discounts for particular customers may, by contrast, appear understandable; for instance, if vendors want to give regular customers a bonus or new customers an incentive for a purchase. In contrast, customers are less likely to accept the orientation on conditions like the operating system that they use or the area where they live. A point to consider is also whether the vendor openly promotes how the price is calculated individually.[96] It is reasonable to argue that disguised spying on customers and manipulating their behavior by vendors may affect the customer's autonomy significantly and therefore automated decisions based on such previous conduct have a similar significant effect on the customers in the sense of Art. 22 (1) GDPR.[97] The Art. 29 Working Party also

---

[87] Schulz (2017), Art. 22 margin note 27.

[88] The European Data Protection Board (2018), the "successor" of the Art. 29 Data Protection Working Party, has endorsed the Guidelines on automated individual decision-making and profiling for the purposes of Regulation 2016/679 17/EN WP251rev.01.

[89] Art. 29 Data Protection Working Party (2018), p. 22.

[90] For more details see Sect. 2.2.4.

[91] Ernst (2017), pp. 1034–1035.

[92] Schulz (2017), Art. 22 margin note 25.

[93] Schulz (2017), Art. 22 margin note 24; Buchner (2018), Art. 22 margin note 26.

[94] Steppe (2017), p. 784.

[95] See Footnote 89.

[96] See Footnote 91.

[97] Malgieri and Comandé (2017) refer in that context also to the EU Unfair Commercial Practices Directive and state in that context that "the protection against pervasive marketing manipulation

considers "the intrusiveness of the profiling process, including the tracking of individuals across different websites, devices and services" as a factor that may lead to the conclusion that the threshold with respect to a similar significant effect is exceeded.[98] However, the example of price discrimination shows that this criterion of "similarly significantly affects him or her" allows for a variety of different interpretations and depending on the interest behind the argumentation can vary widely.

4.1.1.2 Exceptions

As already indicated, there are exceptions to the general rule that automated decision-making must not be used. These are enumerated in Art. 22 (2) GDPR. One of the exemptions is that automated decision-making can be used if it is necessary for entering into, or performance of, a contract between the data subject and a data controller. Automated decision-making is also allowed if the data subject explicitly consented to it or when Union or Member State law, which applies to the controller and which stipulates suitable measures to protect the data subject's rights and freedoms and legitimate interests, authorizes automated decision-making. In the first two cases, the data controller shall according to Art. 22 (3) GDPR implement suitable measures to safeguard the data subject's rights and freedoms and legitimate interests. These shall at least include the right of the concerned individual to involve a human person on the part of the controller to express his or her point of view and to contest the decision.

As noted, the first exception applies when the use of automated decision-making is necessary for the performance of or entering into a contract. The Art. 29 Working Party seems to have taken a rather practical approach and states that the quantity of data can make human involvement impractical or impossible. In such cases automated decision-making may be necessary for contractual or pre-contractual purposes. The example presented by the Art. 29 Working Party is a job recruitment procedure in a very popular company which receives thousands of applications for one advertised position. In such a case, automated filtering may be admissible.[99] A typical example of automated decision-making that is necessary for entering into a contract would be credit rating before granting a loan.[100] Verifying the creditworthiness is required to determine the appropriate loan amount and the level of rates[101] which is, in turn, necessary to prevent defaults and the risk of insolvency.[102] By contrast, the entering into or performance of a contract does not require making a profile of the customers

---

is already legally recognized as a legitimate interest" and argue that "the impairing influence and pervasive manipulation of consumers vulnerability performed through algorithmic decision-making can be considered a 'significant effect' in terms of Article 22(1)," p. 11.

[98] See Footnote 89.

[99] Art. 29 Data Protection Working Party (2018), p. 23.

[100] Schulz (2017), Art. 22 margin note 29–30 and Hladjk (2017), Art. 22 margin note 11.

[101] Schulz (2017), Art. 22 margin note 30.

[102] Hladjk (2017), Art. 22 margin note 11.

preferences and lifestyle, even if the profiling is mentioned in the terms and conditions.[103] The same applies to evaluating how likely it is that a customer will buy a product for a certain price.

Explicit consent may also render the automated decision-making admissible. The general issue about informed consent is that the therein stipulated conditions are not the result of negotiating or dialogue. The content is dictated unilaterally by the controllers.[104] The power asymmetry between the controllers and the data subjects forces the data subjects to either accept the provided conditions or not to make use of the service at all.[105] Data protection friendly services are non-existent or rare, and lock-in effects make the use of these services less attractive, so users often have no choice left but to accept the conditions set up by the service provider or vendor.[106] However, according to Art. 4 No 11 GDPR, consent must be given freely and whether consent given under the described circumstances can be considered as a free decision is questionable.

A further aspect to consider is that the consent must be informed. As indicated earlier, profiling and automated decision-making can be opaque processes. Controllers must make sure that the data subject understands what he or she consents to.[107] This means that controllers must be transparent about what kind of data they process and for what purpose and the consequences of the processing for the user. Unfortunately, users—considering their time, the involved effort and benefit of using the service—often lack the willingness to read the consent declarations critically and, generally, to inform themselves about the privacy intrusiveness of the web services they use.[108] Another important point is that consent declarations often contain only vague descriptions of what kinds of personal data are processed and for what purpose or whether data are transferred to third parties.[109] Besides these general problems of informed consent, specific issues may also have to be considered. In the case of personalized insurance premiums, for example, the future voluntariness of the consent is questioned, as policyholders may feel forced to use personalized rates as other options are not available or too expensive.[110] However, only if no other reasonable standard rates options are available, may the consent be considered as not freely given, whereby it is difficult to determine what is reasonable in that specific context.[111] With regard to price discrimination according to previous shopping behavior of customers, the personalization of prices must be properly explained to customers prior to the processing of the personal data and subsequent decision-making, oth-

---

[103] Art. 29 Data Protection Working Party (2018), p. 13.

[104] Hoffmann-Riem (2017), p. 22.

[105] Hoffmann-Riem (2017), pp. 22–23.

[106] See Footnote 104.

[107] See Footnote 103.

[108] Martini (2017), p. 1019.

[109] See Footnote 104.

[110] Lüdemann et al. (2014), p. 305.

[111] For further remarks on that issue, see Schwichtenberg (2015), p. 380.

erwise, there is no informed consent to justify the processing of personal data and automated decision-making.[112]

As a third option, automated decision-making is also allowed if authorized by Union or Member State law which applies to the controller, and which at the same time stipulates suitable measures to protect the data subject's rights and freedoms and legitimate interests. Recital 71 GDPR mentions as examples laws for fraud and tax evasion monitoring and prevention purposes or to secure the security and reliability of a service provided by the controller. The Union or Member States may restrict the right provided in Art. 22 (1) GDPR as well as on the basis of Art. 23 (1) GDPR.

4.1.1.3 Special Categories of Personal Data

In case special categories of personal data are concerned, Art. 22 (4) GDPR states that decisions based solely on automated processing are only permissible if the conditions laid down in the exemption clauses of Art. 9 (2) (a) or (g) of the Regulation are fulfilled, which means either the data subject must have given explicit consent or a Union or Member State law must legitimize the processing as necessary for reasons of substantial public interest.

4.1.1.4 Suitable Safeguards, Information Obligations and Access Right

4.1.1.4.1 Suitable Safeguards

In those cases where the automated decision-making is legitimated because it is necessary for entering into, or performance of, a contract between the individual and the data controller, or where the data subject gave explicit consent, the data controller is obliged to implement suitable measures to safeguard the data subject's rights and freedoms and legitimate interests.[113] In such cases the concerned data subject has, in accordance with Art. 22 (3) GDPR, at least the right to obtain human intervention on the part of the controller, to express his or her point of view and to contest the decision. Recital 71 GDPR also mentions with respect to automated decision-making that the data subject should have the right "to obtain an explanation of the decision reached after such assessment." The purpose of Art. 22 (3) GDPR is to enable the data subject to ensure that the decision which has been solely based on automated processing will be verified by a human person. This is, of course, only meaningful if the person verifying the decision has the authority and capability to change it.[114] Furthermore, expressing his or her personal view and contesting the decision will only be feasible for the data subject if he or she has a basic understanding of how the decision was taken

[112] Steppe (2017), pp. 777–778.

[113] If the automated individual decision-making is authorized by Union or Member State law, the law must also provide suitable measures to safeguard the data subject's rights and freedoms and legitimate interests (Art. 22 (2) (b) GDPR).

[114] Art. 29 Data Protection Working Party (2018), p. 27.

and on what basis. Whether the data subject can demand an (ex post) explanation of the decision as stated in Recital 71 of the Regulation, is unclear.[115] Art. 22 (3) GDPR itself does not enunciate this specific right in contrast to the "right to obtain human intervention on the part of the controller, to express his or her point of view and to contest the decision," which is explicitly mentioned therein. Furthermore, recitals as such are not legally binding.[116] Looking at the legislative process, it is also evident that the European legislative organs could not reach a consensus to include a right to an explanation in the framework of Art. 22 (3) GDPR. The European Parliament had suggested including a right to an explanation in the actual binding text of the Regulation, but this proposal was not adopted in the trilogue.[117] However, Recital 71 of the Regulation still could be used as a means to interpret what the legislator meant when requiring suitable measures and safeguards to secure the data subject's rights and interests, as long as this does not conflict with the binding legal text or create new rules.[118]

Malgieri and Comandé take the view that this threshold is not exceeded and consequently conclude that Art. 22 (3) GDPR provides a right to obtain an explanation of the decision.[119] Wachter et al., however, make an important point when they argue that Art. 22 (3) GDPR determines clearly by using the formulation "at least" what the minimum obligatory measures and safeguards are.[120] The Art. 29 Working Party—not commenting on the legal effect of Recital 71 GDPR—simply refers to Recital 71 and suggests that "in any case" suitable safeguards should also include "specific information to the data subject…and the right to obtain an explanation of the decision reached after such assessment and to challenge the decision."[121] In this respect, the Art. 29 Working Party emphasizes the need for transparency because the data subjects "will only be able to challenge a decision or express their view if they fully understand how it has been made and on what basis."[122] The detail in which such explanations should be given, is not further explained, but reference is made to the transparency requirements discussed in the context with the data subject's rights under Arts. 13, 14 and 15 of the Regulation.[123] It can be assumed, however, that the main reasons for the decision should be outlined. Otherwise, the data subject cannot express his or her point of view to the decision made. This does not mean that data controllers will have to disclose details about the underlying algorithms or the code

[115]For example, Wachter et al. (2017) deny such a right on the basis of Art. 22 (3) GDPR, pp. 79–80. Malgieri and Comandé (2017) argue for a right to an ex post explanation, pp. 12–13.

[116]Wachter et al. (2017), p. 80.

[117]Wachter et al. (2017), p. 81.

[118]Malgieri and Comandé (2017), pp. 12–13 and Wachter et al. (2017), p. 80.

[119]Malgieri and Comandé (2017), pp. 12–13.

[120]See Footnote 116.

[121]See Footnote 114.

[122]See Footnote 114.

[123]Art. 29 Data Protection Working Party (2018), p. 27; however, this reference is not particularly illuminating for this specific issue as the Art. 29 Working Party seems to understand Art. 13 (2) (f) or Art. 14 (2) (g) GDPR and Art. 15 (1) (h) GDPR as not providing a right to an ex post explanation (see Sect. 4.1.1.4.2).

itself, but they will have to give a basic explanation of the logic of the algorithm and what characteristics have mainly given rise to the decision.[124]

4.1.1.4.2 Information Obligations and Access Right

In this context, the information obligations towards the data subject, regulated in Arts. 13 and 14 GDPR, as well as the data subject's access right, as provided for under Art. 15 GDPR, should also be discussed.[125] While Arts. 13 (1) and 14 (1) GDPR[126] provide basic information obligations, e.g., with respect to the identity of the controller or the purpose of the processing, Arts. 13 (2) and 14 (2) GDPR provide additional information requirements to ensure fair and transparent processing. According to Arts. 13 (2) (f) and 14 (2) (g) GDPR, the data controller shall provide information about the existence of automated decision-making, including profiling, referred to in Art. 22 (1) and (4) GDPR and, at least in those cases, give meaningful information about the logic involved, as well as the significance and the envisaged consequences of such processing for the data subject. The right of access by the data subject under Art. 15 (1) GDPR enables the data subject to request, inter alia, from the controller access to the stored personal data and additional information, e.g., the purpose of the processing. In case automated decision-making in the sense of Art. 22 (1) or (4) GDPR is used by the data controller, the data subject shall, pursuant to Art. 15 (1) (h) GDPR, on his or her request, be informed about the existence of such automated decision-making and shall also be given meaningful information about the logic involved as well as information with regard to the significance and envisaged consequences of such processing.

Looking just at the wording of the provisions in question, the information that should be provided under Arts. 13 (2) (f) or 14 (2) (g) GDPR appears identical to that under Art. 15 (1) (h) GDPR. Nevertheless, it has been argued that the information obligations under Arts. 13(2) (f) and 14 (2) (g) GDPR and the access right regulated in Art. 15 (1) (h) GDPR do not necessarily coincide.[127] Admittedly, there is a consensus that, since the information is to be given before the processing, the obligations under Arts. 13 (2) (f) and 14 (2) (g) GDPR can only concern information about the general system functionality and not about the actual reasons in terms of the future decision.[128] However, commentators have disagreed on whether data subjects may request an ex post explanation of the automated decision (in line with Art. 15 (1) (h)

[124]Datatilsynet, The Norwegian Data Protection Authority (2018), p. 21.

[125]Art. 12 GDPR must be complied with, too. Inter alia, data subjects need to be informed in a "concise, transparent, intelligible and easily accessible form, using clear and plain language…". "The controller shall also facilitate the exercise of the data subject rights under Art. 15 to 22 GDPR…" In general, information shall be provided free of charge.

[126]Art. 13 GDPR applies where the personal data are collected from the data subject. Art. 14 GDPR is relevant if the personal data have not been obtained from the data subject.

[127]Malgieri and Comandé (2017), pp. 13–16.

[128]If pre-defined simplistic or linear models are used, it would be in principle possible to give information about the rationale ex ante (Wachter et al. 2017, p. 79).

GDPR), or whether they can only request the information they (should) have been provided with according to Arts. 13 (2) (f) or 14 (2) (g) GDPR.

This issue is relevant because it is unclear whether Art. 22 (3) GDPR provides an ex post right to an explanation.[129] Thus, whereas Wachter et al. deny a right to an ex post explanation under Art. 15 (1) (h) GDPR,[130] Malgieri and Comandé come to the conclusion that Art. 15 (1) (h) GDPR provides such a right.[131] Interestingly, both found support in the wording of the law and present systematic arguments for their point of view.[132] For its part, the Art. 29 Working Party seems to interpret Arts. 13 (2) (f) and 14 (2) (g) GDPR and Art. 15 (1) (h) GDPR as stipulating congruent information requirements in respect of what information must be provided to the data subject,[133] and consequently denies a right to request "an ex post explanation of a particular decision" under Art. 15 (1) (h) GDPR.[134] From a systematic point of view, as the provisions in question contain identical formulations and, furthermore, considering the legislative history this seems reasonable. In particular, since the European legislator was aware of the issue and merely enshrined a right to obtain an explanation in the non-binding Recital 71 of the Regulation. Hence, the Art. 29 Working Party supports an ex-post right to an explanation under Art. 22 (3) GDPR,[135] but denies it under Arts. 13 (2) (f) and 14 (2) (g) GDPR and Art. 15 (1) (h) GDPR.

With respect to the issue what information must be provided under Arts. 13 (2) (f) or 14 (2) (g) GDPR, and hence under Art. 15 (1) (h) GDPR, the Art. 29 Working Party states that the data controller must inform the data subject "that they are engaging in this type of activity,[136] provide meaningful information about the logic involved, and explain the significance and envisaged consequences of the processing."[137] The controller must explain in a simple manner to the data subject either the rationale

---

[129] See Sect. 4.1.1.4.1.

[130] Wachter et al. (2017), pp. 83–84.

[131] Malgieri and Comandé (2017), pp. 13–14.

[132] For details see Malgieri and Comandé (2017), p. 4 and Wachter et al. (2017), pp. 83–84.

[133] Art. 29 Data Protection Working Party (2018): "Article 15 (1) (h) entitles data subjects to have the same information about solely automated decision making, including profiling, as required under Art. 13 (2) (f) and 14 (2) (g), namely:

- The existence of automated decision making, including profiling;
- Meaningful information about the logic involved, and;
- The significance and envisaged consequences of such processing for the data subject.

The controller should have already given the data subject this information in line with their Article 13 obligations." pp. 26–27.

[134] Art. 29 Data Protection Working Party (2018): "Article 15 (1) (h) says that the controller should provide the data subject with information about the envisaged consequences of the processing, rather than an explanation of a particular decision. Recital 63 clarifies this by stating that every data subject should have the right of access to obtain 'communication' about automatic data processing, including the logic involved, and at least when based on profiling, the consequences of such processing." p. 27.

[135] Section 4.1.1.4.1.

[136] Automated decision-making in the sense of Art. 22 (1) GDPR.

[137] Art. 29 Data Protection Working Party (2018), p. 25.

behind or the criteria which are decisive for the decision (to be taken).[138] Thereby, it is not necessary to provide a complex description of the algorithm or the algorithm itself.[139] However, the explanation must give the data subject an understanding of the reasons for the decision.[140] The Art. 29 Working Party explicitly states that the data subject shall be provided with "general information (notably, on factors taken into account for the decision-making process, and on their respective 'weight' on an aggregate level) which is also useful for him or her to challenge the decision."[141]

4.1.1.4.3 Trade Secrets and Intellectual Property

The impact of trade secrets and intellectual property rights is another issue to consider. The Art. 29 Working Party requires that the data controller must inform data subjects about the relevant factors of the decision as well as "their respective 'weight' on an aggregate level."[142] However, the algorithm as such, the weight of the characteristics used for the probability calculation as well as information about the comparative groups, have been previously evaluated as trade secrets.[143] As such, there is little interest on the side of the controllers to provide information about the algorithm due to the fear of losing their competitive advantage, or that somebody could use the information to "game" the system.[144] This is notwithstanding that, explanations of the system may help users to develop trust.[145]

The obligations on the part of the data controller with respect to creating transparency, as interpreted by the Art. 29 Working Party, and the above interest on the side of the controllers in protecting trade secrets conflict with each other to a certain extent. It remains to be seen in how far data controllers will be able to invoke the protection by trade secrets or intellectual property rights in order to limit their obligations under Arts. 13 (2) (f), 14 (2) (g), 15 (1) (h) and Art. 22 (3) of the Regulation. Recital 63 GDPR states that the right of access (and only that) "should not adversely affect the rights or freedoms of others, including trade secrets or intellectual property and in particular the copyright protecting the software." Nevertheless, it also provides: "However, the result of those considerations should not be a refusal to provide all information to the data subject." It seems appropriate to balance the competing interests. Consequently, data controllers, will not be able to maintain complete silence. They must at least provide specific information about the rationale of the

---

[138]See Footnote 137.

[139]See Footnote 137.

[140]See Footnote 137.

[141]See Footnote 114.

[142]Art. 29 Data Protection Working Party (2018), p. 27. The Norwegian Data Protection Authority also requires that the data subject is informed about "how the data is to be weighted and correlated" (Datatilsynet, The Norwegian Data Protection Authority (2018), p. 21).

[143]BGH (Bundesgerichtshof), judgement from 28 January 2014—VI ZR 156/13; Ernst (2017), p. 1033.

[144]See Footnote 20.

[145]Edwards and Veale (2017), p. 22.

decision[146] and may, for example, use non-disclosure agreements[147] or noisy information in order to protect their interests.[148] Member States may enact corresponding regulations on the basis of Art. 23 (1) (i) GDPR.[149]

4.1.1.4.4 Inscrutable AI

It has already been mentioned above that neural networks, and other forms of machine learning are not wholly transparent, even to their developers. According to the current scientific understanding, the reasoning for a decision generated by such a system cannot be explained precisely.[150] It may, therefore, be impossible to elucidate how information is correlated, and what weight is given to it in a specific decision process.[151] These situations provide particular challenges for the provision of appropriate safeguards as required by Art. 22 (3) GDPR. If the rationale for the decision cannot be discovered by the data controller, it cannot be explained to the relevant subject. In turn, the latter will be unable to contest the decision effectively. It appears then that, where a computer system possesses a certain level of autonomy, data controllers may no longer be able to comply with their information obligations according to Arts. 13 or 14 GDPR, nor the access right under Art. 15 GDPR. These provisions shall, generally, enable the data subject to be informed about which of his or her personal data are processed by the controller and for what purposes and also to make effective use of his or her rights laid down in the GDPR, e.g., Art. 21 GDPR (see Sect. 4.1.3).[152]

4.1.2 General Principles of Data Protection Law

As mentioned earlier, the data processing steps, e.g., the collection of the personal data or the actual profiling, must in any event comply with the general provisions on processing of personal data laid down in the General Data Protection Regulation. This means, for example, that the principles relating to the processing of personal data, regulated in Art. 5 (1) GDPR, must be followed. These principles are: the principle of lawfulness, fairness and transparency; the principle of purpose limitation; the principle of data minimization; the principle of accuracy; the principle of storage limitation and the principle of integrity and confidentiality. The principle of lawfulness

[146]Malgieri and Comandé (2017), p. 22.

[147]See Footnote 146.

[148]Bäcker (2018), Art. 13 margin note 54.

[149]Bäcker (2018), Art. 13 margin note 54; Malgieri and Comandé (2017), p. 22.

[150]McLellan (2016), Knight (2017), Ernst (2017), p. 1027 and Martini (2017), pp. 1018–1019.

[151]Datatilsynet, The Norwegian Data Protection Authority (2018), p. 19.

[152]Schmidt-Wudy (2018), Art. 13 margin note 2; Paal and Hennemann (2018), Art. 13 margin note 4.

has been outlined previously.[153] With regard to the purpose limitation principle, it should be emphasized that for Big Data analytics personal data are often repurposed, which the Regulation though permits only under certain conditions.[154] For now, just a few further remarks with regard to the principle of fairness and transparency shall be made.[155]

First, according to Art. 5 (1) (a) GDPR personal data shall be processed fairly and in a transparent manner. Recital 60 GDPR states that the principle of fairness and transparency requires that the data subject must be informed of the existence of the processing and its purposes as well as "any further information necessary to ensure fair and transparent processing, taking into account the specific circumstances and context in which the personal data are processed." With respect to profiling, it is further stated in Recital 60 GDPR that the data subject should be informed of the existence of profiling and the consequences of such profiling. The extent to which the data subject must be informed about the particulars of the profiling procedure remains unclear. The specific information obligations of Arts. 13 (2) (f) and 14 (2) (g) GDPR apply only in case of automated decision-making in the sense of Art. 22 (1) GDPR.[156] However, where a human person makes the decision on the basis of a profiling result, it may be hard to assess whether a decision has been solely based on automated processing or not. Data subjects may, in any case, have an interest in knowing how their score has been produced, especially as such a score usually has a significant impact for the final decision. The Norwegian Data Protection Authority notes that even though no automated decision-making in the sense of Art. 22 (1) GDPR is involved, it is required by the transparency principle that similar information with regard to the pursued profiling activities is provided to the data subject.[157] The Art. 29 Working Party understands it as good practice to provide the same information as if the requirements of Art. 22 (1) GDPR are applicable.[158]

In accordance with Recital 71 GDPR, the controller should in order to ensure fair and transparent processing in addition use appropriate mathematical or statistical procedures for the profiling. The controller should also implement appropriate technical and organizational measures to avoid inaccurate personal data and to minimize the risk of errors and to prevent discriminatory effects on the basis of racial or ethnic origin, political opinion, religion or beliefs, trade union membership, genetic or health status or sexual orientation. Recital 71 GDPR concerns important key requirements to tackle the challenges posed by profiling and automated decision-making systems. However, as these have been only placed in a non-binding recital, their effect may be marginal. It is also problematic that these requirements may not be sufficiently

[153] Section 4.1 (first paragraph); for more information see Art. 29 Data Protection Working Party (2018), pp. 12–14.

[154] Edwards and Veale (2017), pp. 32–33; detailed elaborations on Big Data and the purpose limitation principle can be found in Forgó, Hänold and Schütze (2017), pp. 17–42.

[155] With regard to the other principles, see Art. 29 Data Protection Working Party (2018), pp. 9–15.

[156] For information on the scope of application of Art. 22 (1) GDPR, see Sect. 4.1.1.1.

[157] Datatilsynet, The Norwegian Data Protection Authority (2018), p. 22.

[158] See Footnote 137.

enforced because the data protection authorities lack the resources required for such undertakings.[159]

4.1.3 Other Relevant Provisions in the General Data Protection Regulation

Another relevant provision with a specific impact on automated decision-making techniques in the sense of Art. 22 GDPR is Art. 35 (1), (3) (a) GDPR. This provides that a data protection impact assessment by the data controller is required in the case of a systematic and extensive evaluation of personal aspects relating to natural persons, which is based on automated processing, including profiling, and on which decisions are based that produce legal effects concerning the natural person or similarly significantly affect him or her.

It should also be mentioned here that the data subject has, according to Art. 21 (1), (2) GDPR to a certain extent the right to object to the processing of personal data for profiling purposes. According to Art. 21 (1) GDPR, the data subject may, on grounds relating to his or her particular situation, at any time object to the processing of personal data concerning him or her which is based on Art. 6 (1) (e) or (f) GDPR, including profiling based on those provisions. Only in case of very compelling reasons may the controller continue with the processing.[160] In case the profiling is done for direct marketing purposes, the data subject has the right to object at any time (Art. 21 (2) GDPR).

4.2 National Laws (Germany)

4.2.1 Section 31 Federal Data Protection Act

In Germany, due to the enactment of the General Data Protection Regulation, the Act to Adapt Data Protection Law to Regulation (EU) 2016/679 and to Implement Directive (EU) 2016/680[161] (Federal Data Protection Act—FDPA) will enter into force on the 25 May 2018. This Act generally only transposes regulatory tasks and opening clauses contained in the GDPR.[162] However, Section 31 of the Act also contains provisions with regard to scoring and credit reports. For example, it specifies further conditions for the use of scoring methods. Inter alia, it is stated in Section 31

---

[159] Weichert (2014) p. 170 and Edwards and Veale (2017), p. 77.

[160] For details, see Art. 21 (1) GDPR.

[161] Gesetz zur Anpassung des Datenschutzrechts an die Verordnung (EU) 2016/679 und zur Umsetzung der Richtlinie (EU) 2016/680 (Datenschutz-Anpassungs- und -Umsetzungsgesetz EU–DSAnpUG-EU)); English version available at: https://www.bvdnet.de/wp-content/uploads/2017/08/BMI_%C3%9Cbersetzung_DSAnpUG-EU_mit_BDSG-neu.pdf. Accessed 17 May 2018.

[162] Greve (2017), p. 737.

(1) of the Act that for the purpose of deciding on the creation, execution or termination of a contractual relationship, the use of scoring shall be permitted only, "if the data used to calculate the probability value are demonstrably essential for calculating the probability of the action on the basis of a scientifically recognized mathematic-statistical procedure." Section 31 of the Act provides other limitations as well, e.g., it prohibits the use solely of address data for scoring purposes. It is questionable how far Member States are allowed to enact such additional provisions.[163] Nevertheless, Section 31 of the Act contains acknowledged principles for scoring which may be considered while applying the actual relevant regulations, e.g., Art. 6 (1) (f) GDPR.[164]

4.2.2 Anti-discrimination Law

Public institutions in Germany have to comply with the principle of equal treatment which is laid down in Art. 3 of the German Constitution[165] (Grundgesetz—GG). According to this principle, all persons shall be equal before the law and no person shall be favoured or disfavoured because of sex, parentage, race, language, homeland and origin, faith, or religious or political opinions. Nor may there be any discrimination based on disability. For the private sector, Art. 3 GG only has limited effect.[166] In general, every person is free to conclude a contract with whomever he or she chooses, and under which conditions.[167]

However, the General Act on Equal Treatment,[168] which implements several European directives,[169] also applies to private persons in certain contexts. The Act prohibits discrimination on the grounds of race or ethnic origin, sex, religion, disability, age or sexual orientation, for example, in professional life as well as in the context of bulk transactions and in the insurance sector. Not every difference of treatment on grounds of the listed criteria is illegal. The law permits a different treatment under certain circumstances, e.g., Section 20 of the Act allows a difference of treatment if

---

[163] Buchner (2018), § 31 BDSG hae notes 4–5; Taeger (2017), pp. 3–9.

[164] Buchner (2018), § 31 BDSG margin notes 6–7.

[165] English version available at: https://www.gesetze-im-internet.de/englisch_gg/. Accessed 17 May 2018.

[166] Bundesverfassungsgericht (Federal Constitutional Court), Decision from 11 April 2018—1 BvR 3080/09.

[167] Bundesverfassungsgericht: "Grundsätzlich gehört es zur Freiheit jeder Person, nach eigenen Präferenzen darüber zu bestimmen, mit wem sie unter welchen Bedingungen Verträge abschließen will" (Decision from 11 April 2018—1 BvR 3080/09).

[168] Allgemeines Gleichbehandlungsgesetz (AGG); English version available at: http://www.antidiskriminierungsstelle.de/SharedDocs/Downloads/DE/publikationen/AGG/agg_in_englischer_Sprache.pdf;jsessionid=1834417026F099B42C9B8BB560277233.2_cid332?__blob=publicationFile&v=3. Accessed 17 May 2018.

[169] See, e.g., Council Directive 2000/43/EC of 29 June 2000 implementing the principle of equal treatment between persons irrespective of racial or ethnic origin, Official Journal L 180, 19/07/2000 p. 0022–0026; Council Directive 2000/78/EC of 27 November 2000 establishing a general framework for equal treatment in employment and occupation, Official Journal L 303, 02/12/2000 p. 0016–0022.

it serves the avoidance of threats or the prevention of damage. Applied to the context of IT-based decision mechanisms, it follows that algorithms—if used in the relevant context—will also need to reflect the above rules. Moreover, algorithms used by public bodies need to comply as well with the more extensive constitutional principle of equality.

It is not possible to legitimize discriminating behavior with the consent of the concerned person (Section 31 of the General Act on Equal Treatment).[170] That is why corresponding agreements, for example, in general terms and conditions would have no legal effect.

5 Discussion

Looking at the current legal provisions for automated decision-making and profiling, it becomes apparent that they suffer from considerable shortcomings when it comes to meeting the challenges reflected in Sect. 3.

One of the identified issues of profiling and automated decision-making techniques, that involve profiling, was that individuals are only perceived as members of a group and not as individuals in their own right. Furthermore, individuals are treated as the passive object of algorithmic evaluation and decision tools and are not able to present their own values and positions. There may also be inaccuracies that result from the automatic application of inference rules by the algorithm in the individual case.[171] The purpose of Art. 22 GDPR is to counteract these risks.[172]

However, as has become apparent from the discussion in this paper, Art. 22 GDPR in reality only achieves a limited protective function. This is because its scope of application covers only decisions solely based on automated processing, but not situations where profiling methods are only used for decision support. As we saw, though, the profiling results often have a crucial impact on the final decisions made by the human decision maker, too.[173] Furthermore, it may be impossible for the concerned individual to verify whether the human decision maker in fact had used his authority and competence to independently review the relevant data or not (issue of automation bias).[174] Another problematic aspect is that the criterion "similarly significantly affects him or her" (which serves to trigger protection for the subject in cases the decision as such has no legal effect) is open to a broad realm of interpretation.[175] The example of price discrimination has shown different ways the criterion might be interpreted, which makes the application of Art. 22 GDPR hard to predict,

---

[170] Ernst (2017), p. 1033.

[171] See Sect. 3.2.

[172] Buchner (2018), Art. 22 margin note 1; Ernst (2017), p. 1030.

[173] Indeed, the European Parliament had proposed to cover also decisions "predominantly" based on automated processing (Wachter et al. 2017, p. 92).

[174] Section 4.1.1.1.1.

[175] Section 4.1.1.1.2.

as long as there is no settled case law.[176] It is also noteworthy, that the exceptions to the prohibition of Art. 22 (1) GDPR are quite wide.[177] Data controllers will—in case the decision-making is not necessary for entering into, or performance of, a contract—be able to rely on the customer's consent to the automated decision-making. This is problematic as these declarations of consent, as previously discussed,[178] are questionably free and informed. Similarly, the stipulation to provide suitable safeguards in Art. 22 (3) GDPR and the information obligations and access right regulated in the GDPR are arguably only of limited help. This will be elaborated on next in the context of how the law meets the needs of transparency with regard to algorithmic processes.

The opaqueness of algorithmic decisions is one of the big concerns associated with profiling and automated decision techniques.[179] The GDPR provides certain instruments to create more transparency for the concerned individual. There are generally fairly comprehensive information obligations on the part of the data controller and the data subject is also entitled to request information with regard to the processing of his or her personal data from the data controller. As previously discussed, Art. 22 (3) GDPR, requiring the implementation of appropriate safeguards, as well as the information obligations of Arts. 13 (2) (f) and 14 (2) (g) GDPR and the access right under Art. 15 (1) (h) GDPR, aim to ensure the data subject receives enough information to understand the rationale of an automated decision, as well as its consequences. Art. 22 (3) GDPR also entitles the data subject to contest the decision.[180]

Nevertheless, as noted above, the protective effect of these regulations is probably limited. Firstly, the scope of these information obligations is disputed[181]; and secondly, these obligations depend as well on the presence of a decision in the sense of Art. 22 (1) GDPR. In case the decision does not fall under Art. 22 (1) GDPR, it is arguable that the data subject, according to the fairness and transparency principle, should be equally informed on that basis.[182] However, as this is not clearly stated in the law, it will probably prove difficult for data subjects to obtain corresponding information about the profiling process, especially because data controllers have little interest in providing such information.[183] In that context, it should also be stressed that it is unclear how far data controllers will be able to invoke the protection by trade secrets or intellectual property rights to limit their obligations under Arts. 13

[176]See Foontoe 176.

[177]Section 4.1.1.2.

[178]See Footnote 177.

[179]Section 3.2.

[180]Section 4.1.1.4.

[181]See Footnote 180.

[182]Section 4.1.2.

[183]Section 4.1.1.4.3; A fitting example is the German credit investigation company SCHUFA which did not want to disclose any information about the comparison groups and the particular weights of the characteristics used for the algorithm to determine creditworthiness (BGH, judgement from 28 January 2014—VI ZR 156/13).

(2) (f), 14 (2) (g), Art. 15 (1) (h) and Art. 22 (3) GDPR.[184] One can imagine that data controllers, for understandable reasons, will try to exploit these limits to their fullest.

Another issue with regard to creating transparency for the concerned persons is that the underlying algorithmic processes can be very complex and, in such cases, it is a difficult task to provide the required information in accordance with Arts. 13 (2) (f), 14 (2) (g), Art. 15 (1) (h) and Art. 22 (3) of the Regulation.[185] For some forms of AI it may even be impossible to explain why the system has made a particular decision or to tell about the rationale behind, or the criteria relied on in reaching the decision. The question is now what the legal implications of these situations will be, considering the purpose of these provisions to give the concerned individual a basic understanding about the underlying process and the rationale of the decision in order to contest it.

Notwithstanding the above, it shall be mentioned here as well that it has been brought forward that transparency alone will not resolve the issue of discriminatory and unfair algorithmic decisions.[186] As mentioned previously, people often lack the willingness to read consent and privacy declarations.[187] It is probable that the same will apply to disclosed information about algorithmic processes.[188] According to Edwards and Veale "individuals are mostly too time-poor, resource-poor, and lacking the necessary expertise" to make use of their rights.[189] In general, the concept of transparency places a heavy burden on people to seek out the necessary information, interpret it and determine their chances to change it, often only to find that they lack the power to effectuate a change.[190] The power disparity may just be too great for the majority of the people to effectively exercise their legal rights against large influential enterprises.[191] This applies all the more when individuals will predominantly deal with algorithmic judgements.[192]

However, more positively, there may rather be an opportunity for consumer groups, academics, or regulatory bodies to make use of more transparent processes in order to exercise pressure on businesses not to use discriminatory or prejudicial algorithms; as yet, though, there are only few such success stories.[193]

Another key point of concern noted above was that profiling and automated decision-making systems often inherit a significant potential for discrimination. As discussed in Sect. 4.2.2, the law prohibits discriminatory behavior to a considerable

---

[184] Section 4.1.1.4.3.

[185] Sections 3.2 and 4.1.1.4.4.

[186] Hacker and Petkovka (2017), p. 16; Edwards and Veale (2017), pp. 65–67; Vedder and Naudts (2017), p. 215.

[187] See Sect. 4.1.1.2.

[188] Hacker and Petkovka (2017), p. 19.

[189] Edwards and Veale (2017), p. 67.

[190] See Edwards and Veale (2017), p. 67.

[191] See also Edwards and Veale (2017), p. 67.

[192] Vedder and Naudts (2017), p. 216.

[193] Hacker and Petkovka (2017), p. 17.

extent, but residual gaps remain. Firstly, the prohibition applies to private stakeholders (nearly) solely in the professional life as well as in the context of bulk transactions and in the insurance sector.[194] Secondly, traditional anti-discrimination law is only to a limited extent able to cope with the new forms of discrimination caused by Big Data processes. The discrimination may be well hidden and scarcely recognizable for humans.[195] In such cases, the individual who is subject to an algorithmic evaluation process will be unable to recognize the discrimination even if he or she has been provided with information about the system functionality and the rationale of the decision. It has been also observed that algorithms form different groups on a wide range of parameters, which are not necessarily linked to the classical types of discrimination, but nevertheless lead to unfair treatment for members of certain groups created by algorithms on the basis of correlations.[196] For these cases traditional anti-discrimination rules offer no solution.

With respect to the privacy issue mentioned in Sect. 3.3, the law provides protection in so far as all profiling steps, which involve the processing of personal data as well as the collection of the personal data, must comply with the General Data Protection Regulation. One major requirement here is the purpose limitation principle, which permits the repurposing of data only under certain circumstances. The principle of lawfulness and the other principles laid down in Art. 5 (1) GDPR also set limits for the processing of personal data. These rules provide in general a fair balance between the interests of the data controllers to process the data and the data subjects in not processing the personal data. The law also provides in Art. 21 (1) GDPR a right to object to profiling which is based on Art. 6 (1) (e) or (f) GDPR, and a similar right in case the profiling is done for direct marketing purposes. The territorial scope of the GDPR has also enlarged in comparison to the legal situation under the Directive. At the same time, the main challenge is and will be to secure enforcement of the data protection provisions. The provisions in the GDPR allowing for higher fines in case of non-compliance, will probably only have a limited effect. Moreover, data controllers may well continue to try and legitimate all processing of personal data using the instrument of consent, which has degenerated into a meaningless tool of (non)self-determination.[197] The issue of "invisible visibility"[198] can hypothetically be met with the access request according to Art. 15 (1) GDPR, though in making such a request individuals must be aware of their right and probably they will not make use of it without a specific cause. The potential to influence people by means of profiling data has seemingly not yet found its way into the minds of most people.

---

[194] It has been suggested to consider expanding the scope of the law by prohibiting discrimination in all cases that are based on algorithmic data assessment (Martini 2017, p. 1021).

[195] Hacker and Petkovka (2017), p. 20.

[196] Vedder and Naudts (2017), p. 217.

[197] See Footnote 187.

[198] See Sect. 3.3.

6 Conclusion

The discussion in this chapter has shown that the increased use of profiling and automated decision-making techniques raises serious challenges for society and that the current legal framework only offers limited solutions. Some of the issues, especially those caused by uncertainty as to how certain provisions are to be interpreted, may be less serious in the future, in the light of guidance in future case law, as well as guidelines from the European Data Protection Board and national data protection authorities available. As regards other issues, e.g., the dilemma of opaque AI, further research in the IT realm might identify possible ways forward. The necessity of understanding algorithmic decisions has been recognized and "explainable AI" has become a hot research topic. However, transparency and individual rights may not suffice. It has been suggested that using algorithmic applications in sensitive fields should be subject to independent control mechanisms, e.g., regarding their code, incorporation of the database and specifications for training processes.[199] In the case of complex machine learning applications or updates, continuous control is arguably necessary.[200] Obligations of secrecy and in-camera-proceedings could protect the interests of the providers.[201] Other authors have identified other possible strategies, e.g., making available an active choice between data-collecting services (paid by data) and data-free services (paid by money).[202] Of course, implementing such strategies will not be possible without first overcoming immense obstacles. However, one thing has become clear: as a society we need to decide whether we want to live in a world that is increasingly determined by algorithms and, if so, under which conditions.

Acknowledgements This work has been supported by the EU project SoBigData (http://www.sobigdata.eu/) which receives funding from the European Union's Horizon 2020 research and innovation program under grant agreement No. 654024 and the German national project ABIDA (http://www.abida.de/) which has been funded by the Bundesministerium für Bildung und Forschung (BMBF). The author would like to thank Marc Stauch and Julia Pfeiffenbring for their valuable support.

References

Article 29 Data Protection Working Party. (2018). *Guidelines on automated individual decision-making and profiling for the purposes of regulation 2016/679*. 17/EN WP251rev.01.

Bäcker, M. (2018). In J. Kühling & B. Buchner (Eds.), *Datenschutz-Grundverordnung BDSG Kommentar* (2nd ed.). Munich: C.H. Beck.

Binns, R. (2017). *How to comply with GDPR Article 22? Automated credit decisions*. Available at: http://webcache.googleusercontent.com/search?q=cache:GDpeS-Ct11AJ:www.reubenbinns.com/blog/how-to-comply-with-gdpr-article-22-automated-credit-decisions/&num=1&client=firefox-b&hl=de&gl=de&strip=1&vwsrc=0. Accessed May 10, 2018.

---

[199] Martini (2017), p. 1021 and Edwards and Veale (2017), pp. 75–77.

[200] Martini (2017), p. 1021.

[201] Martini (2017), p. 102.

[202] Hacker and Petkovka (2017), pp. 2–42.

Buchner, B. (2018). In J. Kühling & B. Buchner (Eds.), *Datenschutz-Grundverordnung BDSG Kommentar* (2nd ed.). Munich: C.H. Beck.

Clifford Chance. (2017). *Me, myself and AI: When AI meets personal data*. Available at: https://talkingtech.cliffordchance.com/en/cybersecurity/me–myself-and-ai–when-ai-meets-personal-data-.html. Accessed May 10, 2018.

Datatilsynet, The Norwegian Data Protection Authority. (2018). *Artificial intelligence and privacy*. Report, January 2018. Available at: https://www.datatilsynet.no/globalassets/global/english/ai-and-privacy.pdf. Accessed May 16, 2018.

Der Tagesspiegel. (2018). *Rekrutierung beim Versicherungskonzern Talanx*. Wo Roboter Manager testen. Available at: https://www.tagesspiegel.de/politik/rekrutierung-beim-versicherungskonzern-talanx-wo-roboter-manager-testen/21147540.html. Accessed 16 May 2018.

Edwards, L., & Veale, M. (2017). Slave to the algorithm? Why a 'right to an explanation' is probably not the remedy you are looking for [draft, August 17, 2017]. *Duke Law & Technology Review, 16*(18), 18–84 (forthcoming).

Ernst, C. (2017). Algorithmische Entscheidungsfindung und personenbezogene Daten. *Juristen-Zeitung, 72*(21), 1026–1036.

European Data Protection Board. (2018). *Endorsement 1/2018*. Available at: https://edpb.europa.eu/sites/edpb/files/files/news/endorsement_of_wp29_documents_en_0.pdf. Accessed June 07, 2018.

Forgó, N., Hänold, S., & Schütze, B. (2017). The principle of purpose limitation and Big Data. In M. Corrales, M. Fenwick, & N. Forgó (Eds.), *New technology, Big Data and the law*. Singapore: Springer.

Greve, H. (2017). Das neue Bundesdatenschutzgesetz. *NVwZ, 36*(11), 737–744.

Hacker, P., & Petkovka, B. (2017). Reining in the big promise of Big Data: Transparency, inequality, and new regulatory frontiers. *Northwestern Journal of Technology and Intellectual Property, 15*(1), 1–42.

Helfrich, M. (2017). In G. Sydow (Ed.), *Europäische Datenschutzgrundverordnung Handkommentar*. Baden-Baden: Nomos.

Hildebrandt, M. (2009). Who is profiling who? Invisible visibility. In S. Gutwirth, et al. (Eds.), *Reinventing data protection?*. Dordrecht: Springer.

Hladjk, J. (2017). In E. Ehmann & M. Selmayr (Eds.), *Datenschutz-Grundverordnung Kommentar*. Munich: C.H. Beck.

Hoffmann-Riem, W. (2017). Verhaltenssteuerung durch Algorithmen—Eine Herausforderung für das Recht. *Archiv des öffentlichen Rechts, 142*(1), 1–42.

Hofmann, K. (2016). Smart factory—Arbeitnehmerdatenschutz in der Industrie 4.0—Datenschutzrechtliche Besonderheiten und Herausforderungen. *ZD, 6*(1), 12–17.

Holzinger, A., et al. (2017). *What do we need to build explainable AI systems for the medical domain?* Available at: http://arxiv.org/pdf/1712.09923v1. Accessed May 10, 2018.

IT Finanzmagazin. (2017). *N26 Echtzeit-Kredit: Per smartphone-app automatisiert zum Kredit bis 25.000 €*. Available at: https://www.it-finanzmagazin.de/n26-echtzeit-kredit-in-wenigen-minuten-per-smartphone-app-automatisiert-zum-kredit-bis-25-000-e-44535/. Accessed May 16, 2018.

James, K. (2015). *6 ways to avoid sneaky online price changes*. WISEBREAD. Available at: http://www.wisebread.com/6-ways-to-avoid-sneaky-online-price-changes. Accessed May 16, 2018.

Jandt, S. (2015). Big Data und die Zukunft des Scoring. *Kommunikation und Recht, 18*(6, Beihefter 2), 6–8.

Kamlah, W. (2003). Das Scoring-Verfahren der SCHUFA. *Multimedia und Recht, 6*(2), V–VII.

Kamlah, W. (2016). In K. U. Plath (Ed.), *BDSG/DSGVO Kommentar zum BDSG und zur DSGVO sowie den Datenschutzbestimmungen von TMG und TKG* (2nd ed.). Cologne: Otto Schmidt.

Knight, W. (2017). The dark secret at the heart of AI. *MIT Technology Review*. Available at: https://www.technologyreview.com/s/604087/the-dark-secret-at-the-heart-of-ai/. Accessed May 10, 2018.

Knoche, K. (2018). *KI: In kritischen Anwendungen muss die künstliche Intelligenz nachvollziehbare Ergebnisse liefern*. IT Finanzmagazin. Available at: https://www.it-finanzmagazin.de/ki-kuenstliche-intelligenz-nachvollziehbare-ergebnisse-67860/. Accessed May 16, 2018.

Kramer, B. (2018). *Der Algorithmus diskriminiert nicht"—Interview*. ZEITONLINE. Available at: https://www.zeit.de/arbeit/2018-01/roboter-recruiting-bewerbungsgespraech-computer-tim-weitzel-wirtschaftsinformatiker. Accessed May 16, 2018.

Kühling, J., & Martini, M. (2016). Die Datenschutz-Grundverordnung: Revolution oder evolution im europäischen und deutschen Datenschutzrecht? *EuZW, 27*(12), 448–454.

Lüdemann, V., Sengstacken, C., & Vogelpohl, K. (2014). Pay as you drive: Datenschutz in der Telematikversicherung. *RDV, 6,* 302–306.

Malgieri, G., & Comandé, G. (2017). Why a right to legibility of automated decision-making exists in the general data protection regulation. *International Data Privacy Law, 7*(4), 243–265.

Martini, M. (2017). Algorithmen als Herausforderung für die Rechtsordnung. *JuristenZeitung, 72*(21), 1017–1025.

Martini, M. (2018). In B. Paal & D. A. Pauly (Eds.), *Datenschutz-Grundverordnung Bundesdatenschutzgesetz* (2nd ed.). Munich: C.H. Beck.

McLellan, C. (2016). *Inside the black box: Understanding AI decision-making*. ZDNet. Available at: https://www.zdnet.com/article/inside-the-black-box-understanding-ai-decision-making/. Accessed May 16, 2018.

Paal, B., & Hennemann, M. (2018). In B. Paal & D. A. Pauly (Eds.), *Datenschutz-Grundverordnung Bundesdatenschutzgesetz* (2nd ed.). Munich: C.H. Beck.

Schaar, P. (2016). *Algorithmentransparenz. ALGORITHMWATCH*. Available at: https://algorithmwatch.org/de/algorithmentransparenz/. Accessed May 25, 2018.

Schermer, B. W. (2011). The limits of privacy in automated profiling and data mining. *Computer Law & Security Review, 27*(1), 45–52.

Schmidt-Wudy, F. (2018). In H. A. Wolff & S. Brink (Eds.), *Beck OK Datenschutzrecht* (23rd ed.). Munich: C.H. Beck.

Schönhaar, L. (2018). *Bewerbung: Unternehmen nutzen eine neue recruiting-Methode, die vor allem jungen und unerfahrenen Bewerbern helfen kann*. Available at: https://www.businessinsider.de/bewerbung-neue-recruiting-methode-koennte-jungen-und-unerfahrenen-bewerben-helfen-2018-2. Accessed May 16, 2018.

Schulz, S. (2017). In P. Gola (Ed.), *Datenschutz-Grundverordnung VO (EU) 2016/679 Kommentar*. Munich: C.H. Beck.

Schwichtenberg, S. (2015). "Pay as you drive"—Neue und altbekannte probleme. *Datenschutz Datensich, 39*(6), 378–382.

Steppe, R. (2017). Online price discrimination and personal data. A general data protection regulation perspective. *Computer Law & Security Review, 33*(6), 768–785.

Stolberg, M., Ceccotti, C. (2018). *XAI—Explainable AI: Wissen was die KI wirklich macht—So bleibt künstliche Intelligenz erklärbar*. IT Finanzmagazin. Available at: https://www.it-finanzmagazin.de/xai-wissen-was-die-ki-wirklich-macht-kuenstliche-intelligenz-erklaerbar-67609/. Accessed May 10, 2018.

Taeger, J. (2017). Verbot des profiling nach Art. 22 DS-GVO und die Regulierung des Scoring ab Mai 2018. *RDV, 33*(1), 3–9.

Vedder, A., & Naudts, L. (2017). Accountability for the use of algorithms in a Big Data environment. *International Review of Law, Computers & Technology, 31*(2), 206–224.

Von Lewinski, K. (2018). In H. A. Wolff & S. Brink (Eds.), *Beck OK Datenschutzrecht* (23rd ed.). Munich: C.H. Beck.

Wachter, S., Mittelstadt, B., & Floridi, L. (2017). Why a right to explanation of automated decision-making does not exist in the general data protection regulation. *International Data Privacy Law, 7*(2), 76–99.

Weichert, T. (2014). Scoring in Zeiten von Big Data. *ZRP, 47*(6), 168–171.

Zuiderveen Borgesius, F., & Poort, J. (2017). Online price discrimination and EU data privacy law. *Journal of Consumer Policy, 40*(3), 347–366.

Artificial Intelligence and Collusion: A Literature Overview

Steven Van Uytsel

Abstract The use of algorithms in pricing strategies has received special attention among competition law scholars. There is an increasing number of scholars who argue that the pricing algorithms, facilitated by increased access to Big Data, could move in the direction of collusive price setting. Though this claim is being made, there are various responses. On the one hand, scholars point out that current artificial intelligence is not yet well-developed to trigger that result. On the other hand, scholars argue that algorithms may have other pricing results rather than collusion. Despite the uncertainty that collusive price could be the result of the use of pricing algorithms, a plethora of scholars are developing views on how to deal with collusive price setting caused by algorithms. The most obvious choice is to work with the legal instruments currently available. Beyond this choice, scholars also suggest constructing a new rule of reason. This rule would allow us to judge whether an algorithm could be used or not. Other scholars focus on developing a test environment. Still other scholars seek solutions outside competition law and elaborate on how privacy regulation or transparency reducing regulation could counteract a collusive outcome. Besides looking at law, there are also scholars arguing that technology will allow us to respond to the excesses of pricing algorithms. It is the purpose of this chapter to give a detailed overview of this debate on algorithms, price setting and competition law.

Keywords Price fixing · Tacit collusion · Conscious parallelism · Rule of reason Per se illegal · Algorithmic collusion · Monitoring algorithm · Parallel algorithm Signaling algorithm · Self-learning algorithms · Reinforcement learning Q-learning · Sandbox texting · White-box testing · Black-box testing

---

S. Van Uytsel (✉)
Graduate School of Law, Kyushu University, Fukuoka, Japan
e-mail: uytsel@law.kyushu-u.ac.jp

M. Corrales et al. (eds.), *Robotics, AI and the Future of Law*, Perspectives in Law, Business and Innovation, https://doi.org/10.1007/978-981-13-2874-9_7

1 Introduction

Ariel Ezrachi and Maurice E. Stucke wrote in their book Virtual Competition, The Promise and Perils of the Algorithmic-Driven Economy that "the upsurge of algorithms, Big Data, and superplatforms will hasten the end of competition law as we know it."[1] In an earlier paper, Artificial Intelligence and Collusion: When Computers Inhibit Competition, both scholars justified this statement by alluding that "when computer algorithms and machines take over the role of market players, the spectrum of possible infringements may go beyond traditional collusion." The idea that machines are taking over the decision making process related to pricing had also been put forward by Salil K. Mehra in earlier works, like the De-Humanizing Antitrust: The Rise of the Machines and the Regulation of Competition[2] and Antitrust and the Robo-Seller: Competition in the Time of Algorithms.[3]

The works of Ezrachi, Stucke and Mehra are thought provoking. Within three years of launching the idea of the "end of competition law as we know it,"[4] several responses have been formulated to the initial scholarship on algorithms and competition law.[5] It is the purpose of this chapter to review the original ideas and the responses and to discern whether there is any pattern within the understanding of the topic.

This chapter is divided in 5 sections. Section 2 will introduce the reader with taxonomy that Ezrachi and Stucke developed to converse about algorithmic collusion and competition law. Some scholars and institutions have given the taxonomy of Ezrachi and Stucke different names. These will be introduced as well. The taxonomy will be followed by the perceptions on whether contemporary competition law and theory is applicable to all the various categories of the taxonomy. Section 3 puts forward different views on the evidence for algorithms colluding. To avoid a "faith-based theory"[6] on algorithms and competition law, this section will expand on the call for empirical evidence, the idea that price discrimination is more likely than collusion, and the need to investigate whether algorithmic homogeneity will be the norm. How to contain algorithms to collude, will be developed in Sect. 4. More specifically, this Section will examine claims that contemporary competition law is flexible enough to apply to most types of algorithmic collusion, at an argument to construct a rule of reason approach towards the use of algorithmic price setting and a call for experimentation. Section 5 concludes.

[1] Ezrachi and Stucke (2016), p. 233.

[2] Mehra (2014).

[3] Mehra (2015).

[4] See Footnote 1.

[5] For a list of references discussing the topic, see infra References.

[6] For an argument in other fields, see, e.g., Lemley (2015), Jacobs (2016).

2 Algorithms and Collusion, a Taxonomy

2.1 *Four Models of Algorithmic Interaction with Pricing Strategies*

The scope of competition law is traditionally divided between agreements and unilateral conduct. Agreements could be further separated into agreements among competitors (horizontal agreements), agreements between market participants that are in a supply relationship with each other (vertical agreements) or concentrations (merger or acquisition). Collusion is an example of an agreement among competitors and is directly or indirectly related to the price the firms will charge for their products or services.

Ezrachi and Stucke contend that collusion "reflects a concurrence of wills between the colluding companies' agents. Illegality is triggered when companies, through their directors, officers, employees, agents, or controlling shareholders, operate in concert to limit or distort competition."[7] A "human centered" cause of collusion may be coming to an end. Pricing algorithms could take over any of the roles traditionally played by the companies' agents. Ezrachi and Stucke divide the role a pricing algorithm can play into 4 categories: messenger, hub and spoke, predicable agent and digital eye.[8,9]

The Secretariat of the Organization for Economic Co-operation and Development (OECD) follows this categorization but uses different names.[10] The OECD distinguished between monitoring algorithms, parallel algorithms, signaling algorithms, and self-learning algorithms.[11] Niccolò Colombo terms these four categories as follows: classical digital cartel, inadvertent hub-and-spoke, tacit algorithmic collusion and dystopian virtual reality.[12]

Messenger (monitoring algorithm or classical digital cartel) embodies a model in which an algorithm is put in place after humans have discussed and agreed to collude. The algorithm's main purpose is to implement, monitor, and police the cartel.[13] In the words of the OECD, this algorithm is to collect "information concerning competitors' business decisions, data screening to look for any potential deviations and eventually the programming of immediate retaliations."[14] Whereas Ezrachi and Stucke exemplify this model with a reference to an airline pricing cartel,[15] Colombo

---

[7]Ezrachi and Stucke (2017a), p. 1782. For a summary and examples, see Schwalbe (2018), pp. 4–6.

[8]Digital eye has also been termed autonomous machine in earlier versions of the Ezrachi and Stucke's work.

[9]Ezrachi and Stucke (2017a) p. 1782. See also Janka and Uhsler (2018), pp. 114–119.

[10]OECD (2017).

[11]OECD (2017), pp. 24–32.

[12]Colombo (2018), pp. 12–14.

[13]Ezrachi and Stucke (2017a), p. 1782.

[14]OECD (2017), p. 24.

[15]Ezrachi and Stucke (2017a), p. 1786.

picks up the poster cartel case.[16] The difference between these two cases is that the algorithm in the former exchanges information to implement a cartel agreement reached by humans,[17] while in the later the algorithm was coded to achieve a fix price for certain posters sold on Amazon Marketplace.[18]

Hub and Spoke (parallel algorithms or inadvertent hub-and-spoke) represents a model in which the same algorithm is used "to determine the market price charged by numerous users."[19] This model is based on the fact that retailers use pricing algorithms provided by the same third-party provider. Colombo adds to this that the retailers do not necessarily have the shared intention to achieve a collusive outcome. This outcome is an inadvertent consequence of the use of a similar pricing algorithm.[20] Though the outcome may be inadvertent, Colombo contends that the firms may have knowledge or at least a presumption that competitors may be using the same pricing algorithm.[21] The OECD Secretariat further explains that this model can be broader than the adoption of the same pricing algorithm. When explaining their parallel algorithm, the OECD Secretariat mentions that parallel algorithms could be instituted by outsourcing the "creation of an algorithms to the same IT companies or programmers."[22] Another example given by the OECD Secretariat is the circumstance in which most companies use a pricing algorithm to follow a market leader, "who in turn would be responsible for programming the dynamic pricing algorithm that fixes prices above the competitive level."[23]

Predictable Agent (signaling algorithm or tacit algorithmic collusion) exemplifies a model in which an algorithm is designed to provide a specific outcome based on the market conditions it can observe.[24] These algorithms are implemented by firms independently from each other. If the algorithms present a similar price as the optimal outcome, this equilibrium resembles that of tacit collusion. In many competition laws, this "equilibrium being established above competitive levels"[25] does not trigger the intervention of competition law. The question is whether this stance is defendable in a digital environment in which the firms increasingly digitize market data to feed their algorithms. Such increased transparency will, especially if similarly designed algorithms are put in place, make more markets vulnerable for tacit collusion. Whereas in a non-digital environment tacit collusion would be limited

---

[16]Colombo (2018).

[17]*United States v. Airline Tariff Publ'g Co., 836 F. Supp. 9 (D.D.C. 1993)*; Ezrachi and Stucke (2017a), p. 1786.

[18]Colombo (2018), p. 12.

[19]Ezrachi and Stucke (2017a), p. 1782.

[20]Colombo (2018), p. 13.

[21]See Footnote 21.

[22]OECD (2017), p. 27.

[23]OECD (2017), p. 27; see also Capobianco and Gonzaga (2017), p. 4.

[24]Ezrachi and Stucke (2017a), p. 1782.

[25]Ezrachi and Stucke (2017a), p. 1793.

to oligopolistic markets, transparency could enlarge tacit collusion to markets with more players.[26]

The most advanced model is the Digital Eye (self-learning algorithms or dystopian virtual reality). This model is based on algorithms that operate with machine learning and deep learning technologies. The algorithm is given a goal to achieve, such as profit maximization. The algorithm itself will then determine the best pricing strategy to obtain that goal.[27] There will be no further human intervention in the price setting process. This model has been described by Ezrachi and Stucke as follows:

> We consider the possibility that the computer developers foresee tacit collusion as one of many possible outcomes – but not necessarily the likeliest outcome. Smart machines may independently optimize profitability by reaching conscious parallelism – or they may not. Note that in this category, the algorithm developers are not necessarily motivated to achieve tacit collusion; nor could they predict when, how long, and how likely it is that the industry-wide use of algorithms would yield tacit collusion. Nor is there any intent or attempt by the developers and user of the algorithm to facilitate conscious parallelism. The firm 'merely' relies on AI.[28]

2.2 *The Taxonomy and the Scope of Competition Law*

The above-described taxonomy has proven extremely useful to identify the possible negative impact of algorithms on consumer welfare. The next step is to seek a linkage between the taxonomy and the scope of competition law. Essential for this quest is to understand to what extent there is an agreement underlying the developed taxonomy. Only two presume the existence of an agreement. Underlying the Messenger is an agreement to fix the price but also an agreement to use a specific computer model to implement that price fixing agreement. Both agreements are put in place by companies' agents. The second model, Hub and Spoke, is based on a vertical agreement between the producer of the pricing algorithm and the company requesting the implementation of the algorithm. Any form of agreement is absent in the last two models. Algorithms are devised independently from each other. What distinguishes the Predicable Agent from the Digital Eye is the intention underlying the algorithm. Algorithms in the former model are conceptualized to respond to competitors' algorithms and this based upon presumptions how the competitors' algorithms will operate. In the latter model, the collusion will be achieved through a "high-speed trial-and-error"[29] learning process.

Due to the existence of an agreement, the Messenger model or monitoring algorithm is the least problematic for competition law. As explained above, this model requires companies' agents to work out a price fixing scheme and reach an agreement to use a computer model to monitor that price fixing scheme. Price fixing agreements

[26] See, e.g., Capobianco and Gonzaga (2017), p. 2.

[27] OECD (2017), p. 30.

[28] Ezrachi and Stucke (2017a), p. 1795.

[29] See Footnote 28.

are in most jurisdictions per se illegal.[30] Characteristic for per se illegality is that the existence of an agreement between companies' agents, or in various jurisdictions even concerted practices, is sufficient to find an infringement of the competition law. Of course, the existence of that agreement or concerted practice must be proven. Whether the implementation of the agreement is facilitated by computer algorithms is not important for the qualification of the infringement. Therefore, and this is not contested within the literature, the Messenger will be within the spectrum of illegality of most competition laws.[31]

Hub and Spoke may also not pose any problem. Within this model, various companies within an industry (spokes) will have a vertical agreement with the same algorithm developer (hub) into deliver the same pricing algorithm to set the price for each of these companies. The vertical agreements as such are not contested under competition law. It is the "parallel use of the same algorithm which may give rise to concerns."[32] In this respect, two elements are contemplated. On the one hand, a study of the algorithm could reveal that it has been designed to facilitate collusion. On the other hand, an investigation into the intent of the companies could disclose that the companies either aimed at reaching collusion or at least knew that collusion could be the result of the use of the algorithm.[33] In many jurisdictions, hub-and-spoke cartels have been declared illegal based on the classical cartel provision.[34]

Neither the Predictable Agent nor Digital Eye require an agreement. Competition agencies will lack enforcement tools to change the market dynamics in these models. In relation to the Predictable Agent, though, Ezrachi and Stucke refer to § 5 of the Federal Trade Act (FTA).[35] The application of this paragraph does not require an agreement, but just an unfair trade practice. For the application of the unfair trade practice, it is said that:

> (1) evidence that defendants tacitly or expressly agreed to a facilitating device to avoid competition, or (2) oppressiveness, such as (a) evidence of defendants' anticompetitive intent or purpose or (b) the absence of an independent, legitimate business reason for defendants' conduct. Accordingly, in Category III, the defendants may be liable if, when developing the algorithms or in seeing the effects, they were: (1) motivated to achieve an anticompetitive outcome, or (2) aware of their actions' natural and probable anticompetitive consequences.[36]

For the Digital Eye, the FTA may even not be a solution. This model is based neither on the existence of an agreement nor the intent to implement a competition restraining tool. Competition agencies will face an "empty enforcement tool kit"[37]

[30] For a comparison between different jurisdictions, see Dabbah and Hawk (2009).

[31] Colombo (2018), p. 12; Ezrachi and Stucke (2017a), p. 1785.

[32] Ezrachi and Stucke (2017a), p. 1788.

[33] See Orbach (2016); some countries, like Japan, have a complex interpretation of their competition law in relation to hub-and-poke type of cartels (and cartel facilitating firms in general), see Kameoka (2014), pp. 44–49.

[34] Dolmans (2017).

[35] Ezrachi and Stucke (2016), p. 68. For a reference how to deal with in the German context, see Janka and Uhsler (2018), pp. 119–120.

[36] Ezrachi and Stucke (2017a), p. 1785.

[37] Ezrachi and Stucke (2016), p. 77.

to tackle the tacit collusion realized by self-leaning technology of algorithms. Being outside the scope of competition law, program designers may have no incentive to diversify the range of algorithms. The similarity among the algorithms may further facilitate mutual understand and "stabilize a collusive outcome."[38] Seen this possible evolution, Ezrachi and Stucke worryingly declare that:

> Here customers are harmed just as much as (if not more than) in our other collusion scenarios (given fewer episodes of retaliation). We therefore witness a new reality: an anticompetitive outcome which we may not readily perceive and with no one to blame. Any reduction in our welfare is 'merely' a side effect of the rise of the machines and their quest to optimize and serve.[39]

The question is whether competition law and theory should accept a *status quo* for the digital environment.[40] Much of the theory applied in contemporary competition law has been developed in a setting lacking transparency. Competition law intervention is partly based on the fact that consumers lack the information on price or product characteristics, disabling him to discover the best deal. The unfolding use of algorithms seems to go parallel with increased data transparency. In the abstract, this should be a welcome evolution.[41] However, taking the predictions of Ezrachi and Stucke into consideration, we are moving towards a transparency paradox. On the one hand, transparency is encouraged to create a competitive environment. On the other hand, increased transparency may facilitate conscious parallelism by algorithms.

3 Seeking Evidence for Conscious Parallelism

3.1 Countering Indirect Evidence with a Turn to Technology

The literature on artificial intelligence and competition law has made some bold claims. Ezrachi and Stucke postulate that "industry's shift to pricing algorithms can spread tacit collusion beyond duopolies to markets with five or six large firms."[42] In his paper on Robo-sellers, Salil K. Merah opined that "automated pricing via algorithmic processing of collected mass data may tend to lead pricing above the competitive level, either via tacit collusion or more robust cartel formation."[43] The OECD takes these claims for granted when it states that "algorithms might affect some characteristics of digital markets to such an extent that tacit collusion could become

---

[38]Ezrachi and Stucke (2016), p. 79.

[39]Ezrachi and Stucke (2017a), p. 1785, Ezrachi and Stucke (2016), p. 79.

[40]It should be noted that several scholars already questioned the theory of tacit collusion. See Posner (1976) (even though he withdrew from his critique, see Posner (2014)) and Kaplow (2013). The early critique has been part of the current scholars debate to indicate that it is not the first time that the theory of tacit collusion has been questioned. See Gal (2018), pp. 27–28.

[41]Ezrachi and Stucke (2017a), p. 1797.

[42]Ezrachi and Stucke (2017b), p. 2.

[43]Mehra (2015), p. 1363.

sustainable in a wider range of circumstances possibly expanding the oligopoly problem to non-oligopolistic market structures."[44] Though these claims exist, European Commissioner Margrethe Vestager indicates that "we certainly shouldn't panic about the way algorithms are affecting the market."[45] Nicholas Petit has interpreted this cautious stance as a call for "evidence-based antitrust policy."[46]

The current literature has little direct empirical evidence that algorithms will eventually lead to collusive strategies. Stucke and Ezrachi substantiate their claim with references to examples of the petroleum industry in Chile, Germany and Perth (Australia).[47] In all three places, the government acted to heighten the transparency of fuel prices. Transparency, so was the idea, would facilitate the consumer to identify lower prices and so align their consumption patterns. The reality, however, was different and this could be called a transparency paradox. Rather than stimulating competition, the increased transparency actually diminished competition and led to even higher prices. In Australia, the industry was able "to develop a stable collusive pricing structure."[48]

Mehra relies on a theoretical model: the Cournot model.[49] The Cournot model is a model to assess oligopolistic behavior of two firms, producing the same good and independently setting the production output. The model suggests that, without communication between the two firms, it is more profitable for each firm to limit the production to a level for which a price above the marginal cost can be asked. Mehra appeals to both Herbert Hovenkamp[50] and Robert Axelrod[51] to further his argument.[52] Hovenkamp is referred to stress that a Cournot model does not offer an incentive to oligopolies to deviate from the supra-competitive price level.[53] Axelrod, on the other hand, has offered, through an experiment of software programs playing a repeated Prisoner's Dilemma, that a colluded outcome will be reached. This will be especially the case if detection of defection can be achieved quickly.[54] In terms of the Cournot duopoly model, this would mean that two firms are repeatedly determining their price while looking at the price setting of each other. In such a situation, a firm will recognize that lowering the price will trigger the same reaction from the other. Lowering the price will only be attractive if the gains from doing so will outperform the cost of losing the supra-competitive interdependent pricing. At this point, Mehra suggests that the repeated Prisoner Dilemma in a Cournot model will aggregate cooperation with the arrival of robo-sellers or algorithms for the following

[44]OECD (2017), p. 35.

[45]Vestager (2017).

[46]Petit (2017a), p. 362.

[47]Stucke and Ezrachi (2017a), pp. 8–13.

[48]Stucke and Ezrachi (2017a), p. 12.

[49]Mehra (2015), p. 1343.

[50]Hovenkamp (2005).

[51]Axelrod (1988).

[52]Mehra (2015), p. 1346.

[53]Hovenkamp (2005), p. 161.

[54]See Footnote 52.

reasons.[55] Algorithms will enable a firm to detect defection from supra-competitive pricing quicker, reducing the potential profit during the first period of a discounted price. The amount of data algorithms can analyze will also enable them to make more accurate decisions and so prevent errors. Similarly, these algorithms will not be subject to biases that could inspire human salesforce to offer promotions.

Because the evidence provided for the claims has been either indirect or anecdotal, Ashwin Ittoo and Nicolas Petit collaborated on a paper investigating the feasibility of algorithms to collude.[56] The paper starts with observing that Bruno Salcedo has already developed a model "in which two firms use algorithms to set prices."[57] Important for Salcedo's model[58] is the assertion that each firm is able to adjust its own algorithm in response to the competitor's algorithm, which it has been able to decode. Even though the algorithm is programmed to price competitively, when the competitor increases the price the algorithm will match the increase. If an algorithm realizes that there is no response from the other algorithm, price increases will be pursued. Once prices are matched frequently, the model concludes that collusion will be the final outcome. The critique of Ittoo and Petit to Salcedo's model is threefold.[59] First, Salcedo's model, designed for a simple form of an oligopoly market, may not deliver the same result in a more complex reality. Second, the assumption of decoding and consequent price matching may not be realistic. Third, Salcedo's model does not take competitive entry into consideration.

To overcome the critique towards Salcedo, Ittoo and Petit revert to Q-learning,[60] which is "a form of model-free reinforcement learning."[61] In other words, Q-learning agents do not have to be constantly reconfigured to reach a specific outcome. These agents can learn what the optimal outcome is by "experiencing the consequences of actions."[62] The learning process is based on trial and error.[63] The Q-learning is thus dynamic.[64]

Within Q-learning, a distinction is made between single-agent Q-learning and multi-agent Q-learning.[65] Single-agent Q-learning algorithms are characterized by the fact that they determine their strategy without "considering other agents' [sic] strategies."[66] The inability of these algorithms to consider the behavior of other agents, makes these algorithms inappropriate to apply in pricing decision in an oligopolistic market. Therefore, the issue of collusion by algorithms needs to be ana-

[55]Mehra (2016), pp. 1344–1351.

[56]Ittoo and Petit (2017).

[57]Ittoo and Petit (2017), p. 4.

[58]Salcedo (2015).

[59]Salcedo (2015), pp. 4–5.

[60]Ittoo and Petit (2017), p. 5; see also Schwalbe (2018), p. 9.

[61]Watkins and Dayan (1992), p. 279.

[62]Salcedo (2015), p. 5.

[63]Ittoo and Petit (2017), p. 5

[64]Deng (2018a), p. 85.

[65]Ittoo and Petit (2017), p. 6.

[66]Ittoo and Petit (2017), p. 7.

lyzed against the background of multi-agent Q-learning technologies.[67] Among the multi-agent Q-learning technologies, so state Ittoo and Petit, is the Nash Q-learning technology the most likely to render a collusive price setting.[68] The MinMax Q-learning technology is excluded because this requires the multiple agents to have opposite goals.[69] Pricing algorithms, it is held, "all pursue a same goal of profit maximization."[70]

Nash Q-learning agents could potentially collude, because the outcome of an algorithm is defined as the best response of one agent to the choice of another agent. Theoretically, the "profit-maximizing Nash Q-learning agents could set their prices in response to each other until a point is reached where no agent has any incentive to deviate from this price level given what it expects others to do."[71] However, there are several limitations on this theoretical preposition. First, firms will only be able to "specify such joint actions and rewards for its own [Reinforced Learning] pricing agent payoffs matrix necessitates access to internal information on competitors that is in principle private."[72] Second, even if the preferred outcome of the algorithm is specified as a Nash equilibrium, there is a data problem.[73] Third, the constant monitoring of the other agents' strategies may invalidate the "convergence properties of Q-learning agents."[74] Fourth, all the afore-mentioned problems will exponentially enlarge with the number of participating firms on the market. The updating process will be extremely time consuming for the algorithm.[75] Fifth, there needs to be a balance between exploiting a specific outcome and exploring whether another outcome would be better.[76] Ai Deng adds to this list of limitations the fact that the current examples operate in an environment in which all parameters are fixed.[77] None of the complex uncertainties as they exist in a real business model are reflected in the models developed by scholars.

In conclusion, Ittoo and Petit state that "algorithmic tacit collusion conjecture should not presently be taken as a given. Significant technological challenges exist that undermine the capabilities of Q-learning algorithms to approach a tacit collusion equilibrium."[78] Deng agrees with Ittoo and Petit.[79] Acknowledging that reinforcement learning is the suitable approach to study the likelihood of algorithms contributing to collusion, Deng contends that the "not all the assumptions underlying some

[67] Ittoo and Petit (2017), pp. 7–10.

[68] Ittoo and Petit (2017), p. 9.

[69] Ittoo and Petit (2017), p. 8.

[70] See Footnote 70.

[71] Ittoo and Petit (2017), p. 9.

[72] Ittoo and Petit (2017), p. 11.

[73] Ittoo and Petit (2017), p. 12.

[74] Ittoo and Petit (2017), pp. 12–13.

[75] Ittoo and Petit (2017), p. 13.

[76] See Footnote 76.

[77] Deng (2018a).

[78] See Footnote 76.

[79] Deng (2018a), p. 86; Deng (2018b), p. 4.

concerns expressed in antitrust scholarship have empirical support."[80] Deng further explains that:

> For example, while algorithms can be designed to tacitly or explicitly collude in relative simple situations such as the classic prisoner's dilemma, as well as in other situations, real-world competitive decisions are much more complex, presenting a non-trivial computational challenge. Second, despite some express concerns, the goal of unilateral profit maximization us unlikely to lead the algorithms into successful tacit collusion. In fact, the research shows technical challenges that make designing algorithms capable of learning to cooperate rather complex.[81]

Despite the above conclusions, Deng states that the research of Adam Lerer and Alexander Peysakhovich is worth noting.[82] These two researchers have shown in the "Coin game" that an algorithm is able to distinguish coins and only pick up the coins that maximizes its profit. Transfer this to an antitrust setting and replace the coins with geographical markets or customers and Deng exemplifies which form of price fixing may be more likely.[83] The research he refers to is remarkable for another reason. Lerer and Peysakhovich managed to create a model in which reinforcement learning algorithms managed to influence other learning algorithms, that opted for a non-cooperative strategy, to deviate from that strategy and start cooperating.[84] All in all, the end conclusion of Deng in relation to this kind of algorithms is clear. Cooperation is here a design feature. Such willfully structured algorithm may be easily within the scope of competition law.[85]

A similar conclusion is arrived at by Ulrich Schwalbe. He states that "collusive behaviour of algorithms is significantly more complex than is suggested in many contributions to that issue."[86] This outcome is the result of an extensive survey of literature on reinforcement learning in the context of game theoretical models. The survey starts with the older literature on reinforcement learning in game theory, to further expand on the findings of computer scientists, and to finally elaborate on reinforcement learning in oligopoly games. Whereas the older game theoretic literature concludes that reinforcement learning does not necessarily end in cooperation,[87] the computer scientists found that algorithms could lead towards cooperative outcomes if the algorithms could send signals to each other.[88] Computer scientists have found that "reinforcement learning like Deep-Q, Model-based Reinforcement Learning, or Win-or-Learn-Fast don't cooperate in a repeated prisoner's dilemma most of the time."[89] The literature on reinforcement learning in oligopoly games has indicated

---

[80] Deng (2018a), p. 85.

[81] See Footnote 81.

[82] See Footnote 81.

[83] See Footnote 81.

[84] See Footnote 81.

[85] Deng (2018a), p. 86.

[86] Schwalbe (2018), p. 23.

[87] Schwalbe (2018), p. 13.

[88] Schwalbe (2018), p. 15.

[89] See Footnote 89.

that cooperation is not necessarily impossible. Whether the use of algorithms results in cooperation will depend on different factors,[90] like the nature of the algorithm,[91] the evolution of the market,[92] or the use of multiple algorithms.[93] Also, the economic literature has focused on communication and its ability to contribute to cooperation. The main lesson of experimental economics is that algorithms' learning how to communicate are still in an embryonic stage. Let it be the ability to communicate that seems to be essential for coordination,[94] it is understandable that "the statement by Ezrachi and Stucke messaging that 'two Artificial Neural Networks and One Nash Equilibrum meet in an online (pub) (sic) hub. After a few milliseconds, a unique silent friendship is formed…' (Ezrachi/Stucke (2017)) does not seem to describe the actual situation."[95]

3.2 *Algorithm Homogeneity*

The literature postulating that the use of algorithms will facilitate tacit collusion assumes "algorithmic homogeneity."[96] Algorithmic homogeneity points to the goal pursued by the algorithm. It is accepted that algorithms will seek profit maximization. This will be characterized by the fact that "all algorithms are programmed to 'monitor price changes and swiftly react to any competitor's price reduction.'"[97] The idea pursued is that each algorithm will move towards the same price in this setting. Though Petit has been the first one to use the term algorithmic homogeneity, Dolmans is one of the first to posit that the pricing algorithm could develop different pricing strategies.[98]

The maximization of long-term profits is most likely the message given to the algorithms determining price setting strategies. In the previous section, we have already indicated that competitive pricing strategies may be more optimal in an environment where data is collected on consumer preferences or where this data has led to the customization of the products. Dolmans points out that, with a separate set of information, the algorithm may eventually choose a predatory pricing or any other exclusionary conduct as an optimum strategy to realize the long-term profit.

---

[90] Schwalbe (2018), pp. 16–17.

[91] Simple algorithms can result quicker in cooperation than algorithms based on deep neural networks. See Schwalbe (2018), p. 17.

[92] Markets are not static. Changes due to market entry or exit, mergers, or innovation could complicate cooperation for algorithms. See Schwalbe (2018), p. 17.

[93] Current experiments mainly look at the self-play of an algorithms and does not take the interplay of different algorithms into consideration. See Schwalbe (2018), p. 17.

[94] Schwalbe (2018), p. 19.

[95] Schwalbe (2018), p. 21 (containing a quote of Ezrachi and Stucke (2017a), p. 1).

[96] Petit (2017a), p. 361.

[97] Geradin (2017), p. 2.

[98] Dolmans (2017), pp. 8–9.

The information may include "rivals' market share, assets, capital reserves, employee count, variable and fixed costs, etc."[99]

Besides the idea that different pricing strategies may be developed by the pricing algorithm, it needs to be investigated whether formulating one goal, i.e. profit maximization, would lead to collusion. Algorithms may need more detailed description to produce a result in one way or another.

Another aspect of algorithmic homogeneity is the spread of a similar algorithm across different firms. This kind of assumption may not be valid. Already Petit, when launching the idea that algorithmic homogeneity is underlying much of the debate on algorithms and collusion, points out that algorithms will be in state of flux. This will either be because the firms operating the algorithm will redesign the algorithms or the algorithms will change based upon their learning processes.[100] The continuing progress in the algorithms means that algorithmic asymmetry will be most likely the reality. Against this reality, competition policy should develop.

3.3 *Price Discrimination More Likely Than Collusion*

Picking up the thread of a technological perspective on algorithms, Deng's discourse on machine learning seems to imply that not all kind of machine learning is currently impossible.[101] Deng categorizes three types of machine learning. Above we have already described the reinforcement learning, which he calls learning through trial and error.[102] Learning through examples[103] and learning through difference[104] are two other types of machine learning. In the context of the former, Deng explains that algorithms, further qualified as artificial neural networks, could manage to predict the preferences of a consumer based upon his past behavior.[105] This will enable firms to "offer personalized product options and associated prices."[106]

Behavioral discrimination has also been identified by Ezrachi and Stucke as a possible outcome of the use of algorithms[107] and has been repeated by several others. Damien Geradin explicitly acknowledges the additional claim of Ezrachi and Stucke.[108] Maurice Dolmans, to the contrary, formulates price discrimination as his critique to Ezrachi and Stucke's claim that tacit price collusion is the likely outcome of increase price transparency, which in turn is caused by the collection of big

[99]Dolmans (2017), p. 8.
[100]Petit (2017a), p. 362.
[101]See Footnote 78.
[102]Deng (2018a), p. 85.
[103]See Footnote 81.
[104]Deng (2018a), p. 84.
[105]See Footnote 105.
[106]See Footnote 105.
[107]Ezrachi and Stucke (2016), pp. 83–131.
[108]Geradin (2017), p 4.

data.[109] He further opines that the information gathered may actually allow for price discounts spread through various digital communication channels.[110] Also, Michal S. Gal, in her paper Algorithms as Illegal Agreements, describes the possibility that collecting data on consumers may contribute to price differentiation.[111] Data will allow firms to create digital profiles of their consumers and use them to engage in personalized pricing.

The predictions about what this all means for the future differ. Dolmans presumes that firms will move in a direction of product customization.[112] Products will be differentiated, eventually by combining them with services. Algorithms will have difficulties to "compare 'like for like' prices"[113] and thus also to "achieve collusive equilibria."[114] Though Gal does not dispute that product customization is one way forward, she does not necessarily agree that it neutralizes the algorithms' game towards colluded prices. Algorithms would enable a "quicker and more accurate multifactored analysis"[115] to see which products are seen as alternatives by the consumers.

Gal portrays different views on what may happen if customization is going to be the future. First, collusion may not be on the price anymore.[116] Market segmentation could become a more rational choice to reap supra-competitive profits from the customers. Second, price collusion may only be an option if the data underlying the personalized pricing can either be shared among the competitors or easily be accumulated by the different competitors.[117] For sure, only algorithms will be able to collude on personalized prices. Third, the outcome of algorithms may neither be personalized pricing nor coordinated pricing, but a price that will be nevertheless higher than a competitive price.[118]

Though Gal attributes several paragraphs to price discrimination, and its potential to prevent collusion, she is pessimistic on the likelihood of it to happen.[119] She makes two observations on the reason:

> First, as Amazon learned the hard way, personalized pricing might create a public backlash. Second, and relatedly, in order to avoid personalized pricing, consumers might prefer to browse anonymously. This, in turn, will limit sellers' ability to engage in targeted advertising. The financial loss from the reduced ability to better identify those potential consumers who

[109]Dolmans (2017), p. 9.
[110]Dolmans (2017), p. 4.
[111]Gal (2018).
[112]See Footnote 110.
[113]See Footnote 110.
[114]See Footnote 110.
[115]Gal (2018), p. 21.
[116]Gal (2018), p. 20.
[117]See Footnote 117.
[118]See Footnote 117.
[119]See Footnote 116.

might buy a product, might well be larger than the loss from not being able to perform personalized pricing. When this is true, personalized pricing will not be practiced.[120]

4 Containing Colluding Algorithms

4.1 The Wide Net of Contemporary Competition Law

Despite the debate on whether algorithms will collude, there is also a literature on how competition law should respond in case algorithms will have a negative effect on the market. When elaborating their taxonomy, Ezrachi and Stucke elaborated how competition law could deal with colluding algorithms. Their argument is that, when no human made agreements exist, contemporary competition law will face problems. Intent has been proposed for the Predictable Agent model.[121] If the enforcement agency cannot rely on an agreement or intent, which is the case in a Digital Eye model, a new stance towards tacit collusion may have to be developed.[122]

The call for looking beyond the concept of agreement has been questioned by Jan Blockx in his paper Antitrust in Digital Markets in the EU: Policing Price Bots.[123] By looking at the competition law practice in the European Union, Blockx argues that intent is not required to establish a competition law infringement. The EU enforcement authority, the Commission, has always emphasized the need to look at "'*expressions*' and '*communications*'"[124] of parties to determine whether there is an infringement of the cartel provision. The intent of the parties has never been taken into consideration, especially not to exempt some companies from their liability. In order for these expressions to be an agreement, Blockx holds, it is sufficient that there has been "an invitation to collude to the other party and that the other party tacitly acquiesces to that invitation."[125] Even in the absence of an invitation, the European Courts have contended that the "communication of commercially sensitive information"[126] from which the other parties have not publicly distanced themselves can be qualified as a competition law infringement. This is commonly known as a concerted practice.[127]

The just-described interpretation the European enforcement authorities have given to its competition law may have important implications for the digital world. Blockx gives an example of this by referring to "multiple competing traders [using] the same supplier for the pricing software and this software improves its performance

[120]See Footnote 116.
[121]See Footnote 117.
[122]Ezrachi and Stucke (2016), pp. 71–81.
[123]Blockx (2017).
[124]Blockx (2017), p. 5.
[125]See Footnote 125.
[126]See Footnote 125.
[127]On concerted practices and computers, see Heinemann and Gebicka (2016).

("learns") using the data obtained from the traders."[128] Another example is the involvement of price trackers embedded in the website of the trader which contractually allow the software to optimize prices of multiple traders."[129] The only gap that may exist in the European competition law exists when the "website is crawled without the consent of its owner and the owner has merely made the pricing information which is crawled public."[130] In terms of the Ezrachi/Stucke taxonomy, Sebastian Felix Janka and Severin Benedict Uhsler argue that the gap is most likely existing in relation to the Predicable Agent and the Digital Eye.[131]

The emphasis on expressions and communications does not take away that intentions are often referred to in European enforcement practice.[132] Assessing conduct in the light of its intention is, however, not done to establish an infringement in the sense of Article 101 of the Treaty of the Functioning of the European Union (TFEU). The intention merely helps establishing the objectives pursued by the conduct. Without any evidence on the intention, the Commission could reach a similar result, be it with a "much more detailed effects analysis."[133] Reverting to intentions is mainly done to facilitate enforcement.

Blockx goes one step further and indicates that EU competition laws imposes on undertakings the obligation to constantly monitor its business relations. This could have important consequences for the debate on algorithms. If an undertaking notices that other undertakings sign vertical agreements with the same algorithm developer, the undertaking should carefully assess its position. Similarly, there rests an obligation on undertakings to monitor its business to make sure it is compliant to the EU competition law provisions.[134] Therefore, undertakings should only operate algorithms that are designed not to collude. For self-learning price algorithms, this obligation means that the undertaking should take the necessary steps terminate collusion from the moment she is aware of the colluded price setting. Blockx further notices that it is well-established practice in EU jurisprudence that developers of algorithms facilitating collusion may be implicated.[135] Colombo suggests that this will not be the case when developers were complying with instructions from the firm requesting the development of a certain algorithm.[136]

If this were not enough to regulate the market in which algorithms operate, the Commission could still intervene without imposing fines. Blockx refers for this purpose to Article 7 of Regulation 1/2003.[137] This Article 7 stipulates that the Com-

---

[128]Blockx (2017), p. 6.

[129]See Footnote 129.

[130]See Footnote 129.

[131]Janka and Uhsler (2018), pp. 120–121.

[132]Blockx (2017), p. 6. See also in the context of Article 102 Treaty on the Functioning of the European Union, Zingales (2018).

[133]Blockx (2017), p. 7.

[134]See also Marty (2017), p. 15.

[135]See Footnote 134.

[136]Colombo (2018), p. 17. See also Schwalbe (2018), p. 22.

[137]Blockx (2017), pp. 9–11.

mission can bring any infringement to an end, "even in the absence of an intention or negligence."[138] The absence of an intention or negligence warrants, based upon a reasoning to the contrary of Article 23 (2) Regulation 1/2003, the fact that the Commission cannot impose a fine. The following scenarios could, according to Blockx, example an intervention without a fine:

> A pricing bot would be considered to be so ambiguous that it may not have been possible for its designer or user to foresee its anticompetitive character, the Commission can prohibit the practice without a fine…if an anticompetitive practice is identified which causes parallel behavior between a number of undertakings but it would be impossible to identify the undertaking which is to blame for the collusion.[139]

4.2 *The Need to Create a Rule of Reason*

Gal's journey along the concepts of competition law is slightly braver than Blockx's approach. Whereas Blockx is focusing on whether intent is a necessary element for the concept of agreement and concerted practice under European competition law,[140] Gal expands the research to check whether the use of advanced algorithms can be an element to transform conscious parallelism into an illegal tacit agreement.[141] Starting from the assumption that conscious parallelism is currently outside the scope of competition law, Gal then describes elements that, when performed together with conscious parallelism, would render such a parallelism into a forbidden tacit agreement. These elements, also called "plus factors," are "circumstantial facts or factors…from which an agreement can be indirectly inferred."[142] One category of plus factors that deserve consideration is the category of facilitating practices. Such practices "are positive, avoidable actions that allow competitors to more easily and effectively achieve coordination by overcoming impediments to coordination, in a way that goes beyond mere interdependence."[143] The question is whether the use of an algorithm could be categorized as a facilitating practice. To use Ezrachi and Stucke's taxonomy, this question should be investigated for the Predictable Agent or Digital Eye model algorithms.[144] There are two further limitations. First, the research can exclude algorithms in which the programmer has consciously incorporated coordinating coding or suppliers have consciously employed such a coded algorithm. The intent of either the programmer or the supplier to engage in the described conduct

[138]Blockx (2017), p. 10.

[139]Blockx (2017) pp. 10–11.

[140]Blockx (2017). See also Blockx (2018).

[141]See Footnote 112.

[142]Gal (2018), p. 29. See also Ballard and Naik (2017), p. 4.

[143]Gal (2018), p. 30. Gal also refers to the current practice of treating facilitating factors as a sub-category of the plus factors. See Gal (2018), p. 31.

[144]As seen above, the other ones could easily fall within the scope of competition law as all of them are linked to one or another form of an agreement.

could constitute the necessary elements for an agreement.[145] Second, the research can neglect algorithms that simply mimic human conscious parallelism. This kind of algorithms would, just like its human equivalent, fall outside the scope of competition law.[146]

To consider whether the use of the remaining algorithms could be qualified as a facilitating practice, and thus as a plus factor, a case-by-case analysis must be made. When engaging in this analysis, Gal emphasizes four issues.[147] First, Gal indicates that not all algorithms are meant to coordinate prices between competitors. Second, if the facilitating effects originates "from the conditions of the digital world – e.g., increased connectivity,"[148] these effects should not be confused with "facilitating effects of using the algorithm."[149] Third, in case the algorithm is combined with other practices facilitating coordination, the assessment should take both the algorithm and the other practices into consideration. Fourth, the analysis should classify the algorithms into algorithms that facilitate coordination among market players on the one hand and competitors on the other hand. Having these issues in mind, Gal develops a rule of reason to separate acceptable from unacceptable algorithms. This rule of reason analysis constitutes of three questions. Depending on the answer of each question, it is determined whether the use of the algorithm should be prohibited or not. The three questions are as follows[150]:

Does the algorithm facilitate or strengthen in a non-negligible way the ability to reach or maintain a jointly profitable market equilibrium?

no → legal

yes ↓

Is the use of the algorithm justified by neutral or procompetitive considerations?

no → illegal

yes ↓

Do these considerations outweigh the algorithm's coordination-facilitating effects, and are the latter needed in order to enjoy the former?

yes → legal

no ↓

illegal

Based upon this theoretical description of the rule of reason analysis, Gal provides five examples of potential problematic use of algorithms:

i. Suppliers consciously use of **similar algorithms even when better algorithms are available to them**. [...].
ii. Firms make conscious use of **similar data** on relevant market conditions **even when better data sources exist**. [...].

[145] Gal (2018), p. 33.
[146] Gal (2018), p. 34.
[147] Gal (2018), p. 38.
[148] See Footnote 148.
[149] See Footnote 148.
[150] Gal (2018), p. 39.

iii. Programmers or users of learning algorithms give them **similar case studies** from which to learn **despite those not being the best case studies readily available**. [...].
iv. Users take actions that make it **easier for their competitors to observe their algorithms and/or their databases**, and their competitors take actions to observe them. [...].
v. The user technologically **"locks" the algorithm** so that it is difficult to change it. [...].[151]

Though Gal has limited the rule of reason approach to a specified group of algorithms, her position should not be understood as relinquishing the need to question contemporary competition law and theory. At the start of her argument to establish the need for a rule of reason analysis for the use of algorithms, Gal engages with the debate of Kaplow and Posner, which centers on "whether classical oligopolistic behavior can be prosecuted as an unlawful agreement."[152]

The rule of reason is a US concept. An efficiency defense under European competition law has to be built on Article 101 (3) TFEU. Colombo claims that such a defense should be considered for price setting algorithms, presuming that these algorithms can generate pro-competitive effects for consumers.[153] Though recognizing that "academic research on the economic impact of algorithmic pricing is relatively limited,"[154] Colombo refers to likely cost reductions triggered by the implementation of the algorithms. These reductions may link to lower search costs for the consumer, to more transparency and so leading to more competition, or to increased production efficiency. The result, so is argued, would be lower prices for the consumer.[155] Cost reduction that is passed on to consumers is, however, not the only requirement for the application of Article 101 (3) TFEU. It is further required that it can be shown that the restriction created by the algorithm creates an overall improvement in production or distribution, the restriction is reasonably necessary to attain the efficiencies and competition should not be totally excluded.

4.3 Algorithmic Consumers Counterbalance Algorithmic Coordination

Algorithms are not only important on the business side. Michal Gal and Niva Elkin-Koren developed the argument that algorithms will also benefit consumers.[156] Algorithms will be able to "make and execute decisions for the consumers by directly

[151] Gal (2018), pp. 41–42 (detailed explanations excluded).
[152] Gal (2018), pp. 27–18.
[153] Colombo (2018), pp. 18–20.
[154] Colombo (2018), p. 19.
[155] See Footnote 155.
[156] Gal and Elkin-Koren (2017).

communicating with other systems through the Internet. The algorithm automatically identifies a need, searches for an optimal purchase, and executes the transaction on behalf of the consumer."[157] Algorithms facilitating the consumers' transactions have been termed "algorithmic consumers,"[158] "digital butlers,"[159] or "digital assistants."[160] Among the algorithmic consumers could include, for example, Amazon's Alexa, Google Home or Apple's HomePod.

Gal argues that the algorithmic consumers could function as a counterbalance for the algorithms used by the suppliers. Her argument is centered on three elements: buyer power, the conceptualization of the decisional parameters and the anonymization of the customer. Algorithmic consumers will represent individual consumers on the market. When the algorithmic consumer groups a large number of users, transactions on the market could become less frequent. The transactions of each individual consumer can now be grouped into one large order by the algorithmic consumer. In such circumstances, suppliers' algorithms may be more inclined to deviate from a coordinated price equilibrium.[161] To circumvent the coordination of the suppliers' algorithms, algorithmic consumers can be coded to "eliminate or at least reduce market failure in the long run."[162] Algorithmic consumers could, for example, be instructed not to buy if coordination is presumed or to apply different purchase strategies. The effectiveness of suppliers' algorithms may depend on their ability to create personalized digital profiles of their customers and "suppliers to increase their profits, by setting the maximum price that each consumer is willing to pay ("personalized pricing")."[163] This kind of price discrimination could be prevented by using an algorithmic consumer as intermediary. An individual consumer's preference will disappear "into one virtual buyer."[164] Gal terms this "anonymization through aggregation."[165]

The biggest critique to the suggestion that algorithmic consumers could constitute a market-based solution for the collusion triggered by the suppliers' algorithms is that several of the algorithmic consumers will not be neutral towards individual consumers. The most popular algorithmic consumers are put on the market by large-scale digital platforms. Each of these platforms, as is shown in the case of the European Union against Google, will have their own agenda.[166] Ezrachi has addressed this issue with a reference to the Jim Carey movie *The Truman Show*, in which "Truman lives an ecosystem in which he was perfectly happy, but it was all a façade. And in

[157]Gal (2017), p. 4.

[158]Gal and Elkin-Koren (2017).

[159]Gal and Elkin Koren (2017), p. 336; see also Stucke and Ezrachi (2016), p. 2.

[160]Stucke and Ezrachi (2016).

[161]Gal (2017), p. 4 (italics added).

[162]See Footnote 158.

[163]See Footnote 158.

[164]Gal and Elkin-Koren (2017), p. 331.

[165]See Footnote 165.

[166]Stucke and Ezrachi (2017b).

the online environment we are not very far from that..."[167] The overreliance on the algorithmic consumer will alienate the individual consumer from the market reality. Individual consumers will not realize that, at the end, they are not getting the best deal imaginable for them.[168]

Neutral counter-algorithms may be required to prevent any of the above-mentioned scenarios. Petit has elaborated significantly on this issue in his presentation Antitrust and Artificial Intelligence: State of Play.[169] Enforcement authorities could stimulate the creation of software counteracting virtual coordination. This software could be based on the information the enforcement authorities gather when they are auditing and testing the suppliers' algorithms.[170] Less intrusive market intervention would be if the enforcement authority would specify standards which should be included in the suppliers' algorithms[171] or through popups warning of behavioral discrimination on a website.[172] An alternative would be that the authority just takes a cooperative role and would work with Standard Setting Organizations to formulate the antitrust standards.[173] Active participation in the market, for example through the release of lower prices that could trigger price wars or through instant messages to consumers informing that the platform is not offering the lowest price, is discussed as under the heading of "'digital half' of the competition agency."[174] Petit's list of alternative solutions to the algorithmic consumer, is further supplemented by Ezrachi and Stucke. They point out that a neutral algorithm could be delivered by a consumer cooperative. If required, the market entrance of this cooperative should be stimulated by the government.[175]

4.4 *Auditing or Sandbox Testing the Algorithm*

Unlike with real world price fixing schemes, the evidence of algorithmic collusion is something that is readily available at the premises of the firms using it. To prevent that the algorithms would engage in collusive practices, the algorithms could either be audited or taken into a sandbox and tested. Neither of these suggestions seem to be viable for Ezrachi and Stucke.[176] Auditing the algorithm is a feasible option

---

[167] UNCTAD (2016). See also Ioannidou (2018), p. 9.

[168] For a detailed analysis of the problem that algorithmic consumers may cause, see Ioannidou (2018).

[169] Petit (2017b).

[170] Petit (2017b), slide 15.

[171] Petit (2017b), slide 16.

[172] Petit (2017b), slide 17.

[173] Petit (2017b), slide 16. The appeal to soft law has also been made by Colombo. See Colombo (2018), p. 21.

[174] Petit (2017b), slide 18.

[175] Ezrachi and Stucke (2016), pp. 228–229.

[176] Ezrachi and Stucke (2016), pp. 230–231.

for algorithms that are coded to collude.[177] Not all algorithms, especially not the self-learning ones, will reveal that they, sooner or later, will engage in collusion. Sandbox testing will be an artificial environment in which the algorithm is tested. It is not sure that the algorithm will render a collusive outcome in the sandbox.[178] Equally, the collusive outcome in a sandbox is not necessarily the outcome that will be achieved in a complex real world situation, in which the industry is developing new standards at a rapid pace.[179]

The literature on testing the algorithm has been given a new impulse by the article of Joseph Harrington, titled Developing Competition Law for Collusion by Autonomous Artificial Agents. Harrington surveys the US competition law to state that it requires "a common understanding among firms that they will restrict competition in some manner." To proof the restriction, the enforcement authorities require "express and direct communication that conveys a plan to coordinate behavior." But, the "communication need not to be so egregious." As has been mentioned-above, plus factors could indicate that parallelism is unnatural and should therefore be punished.[180]

Against this background, Harrington identifies four different views on algorithmic collusion. First, there will be algorithms determining the price based upon "information that would be present under competition, such as past prices, sales, and other market data." It is well possible that, based upon the information, the prices will the same or similar. Absent any form of communication, it is hard to argue that competition law should apply. Even if the evidentiary standard changes, Harrington claims that the firms using the algorithms will not be liable. At the end, managers were acting independently and could not foresee that collusion would be achieved. Second, there will be algorithms reaching a collusive outcome because of their coding. Either by examining the algorithm or by testing through the feeding of data to the algorithm, enforcement authorities will be able to determine whether the employed algorithm is illegal. This kind of algorithms should be per se illegal. Third, learning algorithms that seek to enhance efficiency will, in principle, not lead to collusive outcomes. The reason for this claim is that the processing of information[181] for setting the price will fall short of what is necessary to achieve collusion. Fourth, despite the belief in competitive outcomes, there could be cases in which learning algorithms,

---

[177]Ezrachi and Stucke (2016), p. 230. See also Oxera (2017), p. 30.

[178]Ezrachi and Stucke (2016), p. 231. There seems to be discussion on the feasibility or desirability to engage in auditing. See Colombo (2018), p. 20.

[179]Ezrachi and Stucke (2016), p. 231.

[180]See above Sect. 4.2 The Need to Create a Rule of Reason.

[181]The information that the algorithm will take into consideration is most likely past data of the firm to analyze the relationship between price and profit. This data could be linked to current data on the market conditions. Another source of information could be Big Data, information gathered on from consumers, from sales, or even from rival's firms.

more in specific estimation-optimization algorithms[182] and reinforcement learning algorithms,[183] present colluded outcomes.

To determine whether prohibited algorithms have been used, Harrington suggests two testing approaches: "white-box testing" and "black-box testing." The former requires access to the coding,[184] while the latter only permits the tester to observe the input and the output.[185] The problem with white-box testing is that it will only be applicable to algorithms that are decodable. As suggested by Schwalbe, algorithms will likely only achieve coordination if the algorithms can send signals to each other. This requires, with the current state of the art, communication protocols within the algorithms.[186] White-box testing will be problematic with deep learning. Inspection of the code will not reveal any information on whether collusion could be achieved. Therefore, the black-box testing is suggested as alternative. Black-box testing requires the user to feed the algorithms with information on the market condition and how prices respond to these conditions. There are critiques to this black-box testing. First, to get an accurate outcome of the black-box testing, a large set of information may be required. Second, as these algorithms keep on learning, the outcome at the stage of testing may be different than when the algorithm is or was operating in the market.[187]

To overcome the problems identified above, Harrington suggest setting up a research program "for restricting AAs [artificial agents] not to collude, and detecting them when they collude." The research program should be conducted along three steps: first, a collusion incubator should run tests with algorithms to identify when it produces collusion or competition as outcome; second, identify the properties present when a colluded outcome is achieved and this by comparing with the properties of algorithms leading up to competitive pricing; third, re-test the algorithms that have been instructed not to select certain pricing strategies. Since this kind of research program still has to be undertaken, Harrington is not able to state what kind of properties will be identified as problematic. Yet, he guesses that one of the properties that should be forbidden in an algorithm is that it should not match the rival firm's price. Harrington stresses that his test should not lead to the prohibition of price matching, but to the ban on algorithms that result in price matching. Harrington also

---

[182]Defined as "An estimation optimization algorithm estimates the environment faced by a firm and then determines what conduct performs best for that estimated environment. It can deliver a forecast on performance (e.g., profit or revenue) for any action (e.g., price) or strategy (e.g., pricing algorithm). An estimation-optimization algorithm learns over both the environment and the best action for an environment."

[183]Defined as "reinforcement learning fuses these two learning processes by learning directly over actions (or strategies); it figures out what action (or strategy) is best based on how various actions (or strategies) have performed in the past. It does not explicitly estimate the firm's environment (e.g., it does not estimate the firm's demand function) and thus is seen as "model free" because it is not based on a particular model of the firm's environment".

[184]See Footnote 193.

[185]See Footnote 193.

[186]Schwalbe (2018), p. 15.

[187]See also Ezrachi and Stucke (2016), pp. 230–231.

indicates that one should investigate to exclude algorithms that are conditioned to act in response to the rival firm's price setting.

The main critique of Harrington's approach is that the auditing will be a "gargantuan task."[188] There will be various algorithms in use. These algorithms will constantly evolve, either by programmers or by self-learning. To avoid continuous examination of algorithms, Schwalbe suggests to counter the problem by coding. At the fundamental level, each pricing algorithm should be coded so that coordination is less likely to happen.[189]

4.5 *Enhancing Privacy and Reducing Transparency*

The collection of data is going to play a vital role in the use of price setting algorithms. The data could be linked to an individual consumer and so constitute her digital profile. Whenever an algorithm recognizes the digital profile, the algorithm could adjust its price setting according to what it can predict about the consumer based upon what it knows. To avoid the algorithm to link us with a specific digital profile, there may be a turn to anonymous browsing.[190]

An alternative to anonymous browsing would be to increase the privacy of the consumers. Ezrachi and Stucke see two viable solutions.[191] On the one hand, these scholars suggest that consumers should be familiarized with the business practices in the digital environment. Increased awareness could be achieved by requesting the websites drawing information from our digital profile for more openness. Several examples of how this could be done are given. Ezrachi and Stucke suggest that pop-up windows could warn the consumer when information is being gathered or used, websites should provide information on claims being made in relation to the price or the availability of products, or a website could reveal that a personalized price is displayed.[192] On the other hand, the scholars opine that a legal interference may be necessary to align the Internet operators with the privacy needed for consumers. Ezrachi and Stucke point out that Europe seems to move in the latter direction with the adoption of the General Data Protection Regulation 2016/679 (GDPR). The GDPR gives the consumers more control over their data.[193]

If there is preparedness to tackle privacy issues, the intervention could focus on reducing price transparency. This intervention focusses on the seller's side rather than on the consumer's side. One possible solution would to encourage firms to allow secretly communicate with buyers. Governments may also intervene to reduce

---

[188]Schwalbe (2018), p. 23. See also Marty (2017), p. 15.

[189]Schwalbe (2018), p. 23.

[190]See Footnote 116.

[191]Ezrachi and Stucke (2016), pp. 226–228.

[192]Ezrachi and Stucke (2016), p. 227.

[193]Ezrachi and Stucke (2016), p. 227. See also Dolmans (2017), p. 20.

the speed of adjusting prices. To allow for price reductions, the prohibition to swiftly adjust the price could only apply to price increases.[194]

5 Conclusion

A call for rethinking tacit collusion has been launched. The basis for this call is the predicted change algorithms will bring to price setting. It is prophesied that algorithms, together with the gathering of Big Data, will – on the one hand – increase the speed with which tacit collusion could be achieved and – on the other hand – enlarge the market scope in which tacit collusion could be realized. The main question in this debate is whether contemporary competition law can be applied to all scenarios in which algorithms set the price.

This question received an added new dimension when Ezrachi and Stucke developed their taxonomy to discuss algorithms and collusion. Their expose on the topic also triggered others to write on the issue, creating two different lines of thought. On the one hand, there is a discussion on evidence of whether algorithms can collude tacitly. Based upon the empirical evidence, we should be rather pessimistic for now. Technology is not yet well-developed to let computers successfully collude without human intervention. That does not mean, however, that this could not change in the future. Further, the literature also posits that the collusion may be more difficult by attacking the assumption of algorithm homogeneity that is underlying much of the literature.

On the other hand, there is a line of argument that, on a more faith-based approach, algorithms may not necessarily evolve towards collusion. This part of the debate either suggests that algorithms may facilitate discriminatory pricing behavior or, in the worst case, result in other anti-competitive pricing strategies.

Presuming that algorithmic collusion can or may occur, a diverse set of solutions has been suggested. The most conservative one is to argue that the current law is broad enough to cover the technological evolution of algorithmic collusion. If this approach did not allow hard enforcement, other warning systems, sometimes backed up with fines, could be relied on. Others suggest developing a special rule of reason or a system to audit the algorithms. An alternative approach would be to enhance the privacy of consumers or to reduce price transparency, both with the aim of disabling systems to exploit their advantage in the market.

Whichever the evolution will be in the future, the literature shows that it is developing in different directions. Even though the empirical evidence suggest that collusion is not likely, there is an agreement that artificial intelligence will progress. It is in preparation of such an event that the literature should develop possible ways of dealing with the technological progress.

---

[194]Ezrachi and Stucke (2016), pp. 229–230.

References

Axelrod, R. (1988). The evolution of cooperation. In A. A. Gromyko & M. E. Hellman (Eds.), *Breakthrough: emerging new thinking* (pp. 185–192) (adapted from Axelrod, The Evolution of Cooperation. New York: Basic Books, 1984).

Ballard, D. I., & Naik, A. S. (2017). Algorithms, artificial intelligence, and joint conduct. *Competition Policy - International Antitrust Chronicle, 1*(2), 29.

Blockx, J. (2017). Antitrust in digital markets in the EU: Policing price bots. In *Radbound Economic Law Conference*. https://papers.ssrn.com/sol3/papers.cfm?abstract_id=2987705. Accessed July 1, 2018.

Blockx, J. (2018). Policing price bots: algorithms and collusion. In *ASCOLA Conference*. http://www.law.nyu.edu/sites/default/files/upload_documents/Blockx.pdf. Accessed July 1, 2018.

Capobianco, A., & Gonzaga, P. (2017). Algorithms and competition: Friends or foes. *Competition Policy International - Antitrust Chronicle, 1*(2), 1–6. https://www.competitionpolicyinternational.com/wp-content/uploads/2017/08/CPI-Capobianco-Gonzaga.pdf. Accessed July 1, 2018.

Colombo, N. (2018). Virtual competition: Human liability vis-à-vis artificial intelligence's anticompetitive behaviours. *European Competition and Regulatory Law Review, 2*(1), 11–23.

Dabbah, M. M., & Hawk, B. E. (2009). *Anti-Cartel enforcement worldwide (volumes I–III)*. Cambridge: Cambridge University Press.

Deng, A. (2018a). An antitrust lawyer's guide to machine learning. *Antitrust, 32*(2), 82–87.

Deng, A. (2018b). *What do we know about Algorithmic Tacit Collusion*. https://papers.ssrn.com/sol3/papers.cfm?abstract_id=3171315. Accessed July 1, 2018.

Dolmans, M. (2017). Artificial intelligence and the future of competition law—further thoughts (reaction to Prof. Ariel Ezrachi), *GCLC Lunch Talk: Algorithms and Markets: Virtual or Virtuous Competition?* https://www.coleurope.eu/sites/default/files/uploads/event/dolmans.pdf. Accessed July 1, 2018.

Ezrachi, A., & Stucke, M. E. (2016). *Virtual competition: The promise and perils of the algorithm-driven economy*. Cambridge: Harvard University Press.

Ezrachi, A., & Stucke, M. E. (2017a). Artificial intelligence & collusion: When computers inhibit competition. *University of Illinois Law Review, 2017*(5), 1776–1808.

Ezrachi, A., & Stucke, M. E. (2017b) Algorithmic collusion: Problems and counter-measures. OECD Roundtable on Algorithms and Collusion (21–23 June 2017), DAF/COMP/WD (2017) https://www.oecd.org/officialdocuments/publicdisplaydocumentpdf/?cote=DAF/COMP/WD%282017%2925&docLanguage=En. Accessed July 1, 2018.

Gal, M. S., & Elkin-Koren, N. (2017). Algorithmic consumers. *Harvard Journal of Law & Technology, 30*(2), 1–44.

Gal, M. S. (2017). Algorithmic-facilitated coordination: Market and legal solutions. *Competition Policy International—Antitrust Chronicle, 1*(2), 22–28.

Gal, M. S. (2018). Algorithms as illegal agreements. *Berkeley Technology Law Journal* (Forthcoming). https://papers.ssrn.com/sol3/papers.cfm?abstract_id=3171977. Accessed July 1, 2018.

Geradin, D. (2017). Algorithmic tacit collusion and individualized pricing: Are antitrust concerns justified? In *Copenhagen Economics Conference*. https://www.copenhageneconomics.com/dyn/resources/Filelibrary/file/6/66/1498204706/geradin.pdf. Accessed July 1, 2018.

Heinemann, A., & Gebicka, A. (2016). Can Computers form Cartels? About the need for European institutions to revise the concertation doctrine in the information age. *Journal of European Competition Law and Practice, 7*(7), 431–441.

Hovenkamp, H. (2005). *Federal Antitrust Policy: The law of competition and its practices* (3rd ed.). Saint-Paul: Thomson/West.

Ioannidou, M. (2018). Digital agoraphobia: An enforcement perspective. In *2018 ASCOLA Conference*. http://www.law.nyu.edu/sites/default/files/upload_documents/Ioannidou_0.pdf. Accessed July 1, 2018.

Ittoo, A., & Petit, N. (2017). Algorithmic pricing agents and tacit collusion: A technological perspective. https://papers.ssrn.com/sol3/papers.cfm?abstract_id=3046405. Accessed July 1, 2018.

Jacobs, M. J. (2016). Keynote address: Faith-based intellectual technology, faith-based antitrust? In *Keynote Address at the 12th Annual Asian Competition Forum*. https://asiancompetitionforum.com/2016-conference-materials. Accessed July 1, 2018.
Janka, S. F., & Uhsler, S. B. (2018). Antitrust 4.0—The rise of artificial intelligence and emerging challenges to antitrust law. *European Competition Law Review, 39*(3), 112–123.
Kameoka, E. (2014). *Competition law and policy in Japan and the EU*. Cheltenham: Edward Elgar Publishing.
Kaplow, L. (2013). *Competition policy and price fixing*. Princeton: Princeton University Press.
Lemley, M. A. (2015). Faith-based intellectual property. *UCLA Law Review, 62,* 1328–1346.
Marty, F. (2017). *Algorithmes de Prix*. Intelligence Artificielle et Equilibres Collusifs, Science Po OFCE Working Paper No. 14. https://www.ofce.sciences-po.fr/pdf/dtravail/WP2017-14.pdf. Accessed July 1, 2018.
Mehra, S. K. (2014). *De-humanizing antitrust: The rise of the machines and the regulation of competition*. Temple University Legal Studies Research Paper No. 2014-43. https://papers.ssrn.com/sol3/papers.cfm?abstract_id=2490651. Accessed July 1, 2018.
Mehra, S. K. (2015). *Antitrust and the robo-seller: Competition in the time of algorithms*. Temple University Legal Studies Research Paper Series No. 2015-15. https://papers.ssrn.com/sol3/papers.cfm?abstract_id=2576341. Accessed July 1, 2018.
Mehra, S. K. (2016). Antitrust and the robo-seller: Competition in the time of algorithms. *Minnesota Law Review, 100,* 1323–1375.
OECD. (2017). *Algorithms and collusion—Background note by the secretariat*. DAF/Comp(2017)4. https://one.oecd.org/document/DAF/COMP(2017)4/en/pdf. Accessed July 1, 2018.
Orbach, B. (2016). *Hub-and-spoke conspiracies*. Arizona Legal Studies Discussion Paper No. 16-11. https://papers.ssrn.com/sol3/papers.cfm?abstract_id=2765476. Accessed July 1, 2018.
Oxera Consulting LLP. (2017). *When algorithms set prices: Winners and losers*. Discussion Paper. https://www.regulation.org.uk/library/2017-Oxera-When_algorithms_set_prices-winners_and_losers.pdf. Accessed July 1, 2018.
Petit, N. (2017a). Editorial: Antitrust and artificial intelligence: A research agenda. *Journal of European Competition Law and Practice, 8*(6), 361–362.
Petit, N. (2017a). *Antitrust and artificial intelligence: State of play, gclc lunch talk: Algorithms and markets: Virtual or virtuous competition?* https://www.coleurope.eu/sites/default/files/uploads/event/petit_0.pdf. Accessed July 1, 2018.
Posner, R. (1976). *Antitrust law: An economic perspective*. Chicago: The University of Chicago Press.
Posner, R. (2014). Review of Kaplow, competition policy and price fixing. *Antitrust Law Journal, 79,* 761–768.
Salcedo, B. (2015). *Pricing algorithms and tacit collusion*. http://brunosalcedo.com/docs/collusion.pdf. Accessed July 1, 2018.
Schwalbe, U. (2018). *Algorithms, machine learning, and collusion*. https://www.uni-hohenheim.de/qisserver/rds?state=medialoader&objectid=11011&application=lsf. Accessed July 1, 2018.
Stucke, M. E., & Ezrachi, A. (2016). *Is your digital assistant devious?* University of Tennessee Legal Studies Research Paper No. 304. https://papers.ssrn.com/sol3/papers.cfm?abstract_id=2828117. Accessed July 1, 2018.
Stucke, M. E., & Ezrachi, A. (2017a). *Two artificial neural networks meet in an online hub and change the future (of competition, market dynamics and society)*. Legal Studies Research Paper Series No. 323. https://papers.ssrn.com/sol3/papers.cfm?abstract_id=2949434. Accessed July 1, 2018.
Stucke, M. E., & Ezrachi, A. (2017b). *Looking up in the data-driven economy*. University of Tennessee Legal Studies Research Paper No. 333. https://papers.ssrn.com/sol3/papers.cfm?abstract_id=2975510. Accessed July 1, 2018.
United Nations Conference on Trade and Development (UNCTAD). (2016). Q&A with Ariel Ezrachi, Professor of Competition Law, Oxford. http://unctad.org/en/pages/newsdetails.aspx?OriginalVersionID=1362. Accessed July 1, 2018.

Vestager, M. (2017). Algorithms and competition. In Bundeskartellamt 18th Conference on Competition. https://ec.europa.eu/commission/commissioners/2014-2019/vestager/announcements/bundeskartellamt-18th-conference-competition-berlin-16-march-2017_en. Accessed July 1, 2018.
Watkins, C. J. C. H, & Dayan, P. (1992). Q-learning. *Machine Learning, 8*, 279–292. http://www.gatsby.ucl.ac.uk/~dayan/papers/cjch.pdf. Accessed July 1, 2018.
Zingales, N. (2018). Antitrust in the age of algorithmic nudging. In *2018 ASCOLA Conference*. http://www.law.nyu.edu/sites/default/files/upload_documents/Zingales.pdf. Accessed July 1, 2018.

Taming Artificial Intelligence: "Bots," the GDPR and Regulatory Approaches

Sam Wrigley

Abstract Bots and AI have the potential to revolutionize the way that personal data is processed. Unlike processing performed by traditional methods, they have an unprecedented ability (and patience) to gather, analyze and combine information. However, the introduction of "smarter" computers does not always mean that the nature of the processing will change; often, the result will be substantially similar to processing by a human. We cannot, then, regulate processing by bots and AI as a sui generis concept. This chapter examines the different regulatory approaches that exist under the new General Data Protection Regulation (the GDPR)—the general regulatory approach (which treats all processing in the same way), the specific regulatory approach (which imposes specific rules for automated processing) and the co-regulatory approach (where data controllers are required to analyze and mitigate the risks on their own). It then considers how these approaches interact and makes some recommendations for how they should be interpreted and implemented in the future.

Keywords Bots · AI · Data protection · Regulatory methods · GDPR

1 Introduction

AI and bots have the potential to revolutionize the processing of personal data. Technology is advancing to the point where a computer program can independently perform the entire scope of processing activities. An AI can, independently, search for information, decide how to process it, perform that processing, and then carry out an action on the basis of those results. This is sometimes seen as an impossible challenge for law. One could be forgiven for worrying about how to regulate science-fiction's concept of a true, human-like AI—a computer which has a tireless ability to gather and analyze information without any guidance or control from humans. One may be

S. Wrigley (✉)
Faculty of Law, University of Helsinki, Helsinki, Finland
e-mail: sam@thewrigleys.com

M. Corrales et al. (eds.), *Robotics, AI and the Future of Law*, Perspectives in Law, Business and Innovation, https://doi.org/10.1007/978-981-13-2874-9_8

even be forgiven for conceiving of such an AI as an unknowable being, capable of making decisions that nobody else could reach or understand, beyond all constrains of morality, emotion or humanity, and which will process data in seemingly monstrous ways that significantly prejudice data subject rights. Attempting to regulate such an intelligence would, indeed, be a herculean task. In addition to asking what rules we would need to place upon AI, a fundamental challenge would simply be how to ensure that the legislation targeted it correctly.

Fortunately, the task of regulating this AI is not one with which the law must concern itself. To begin with, such an AI does not exist in modern technology. While bots and AI are becoming increasingly capable and intelligent, we are still a long way from a true, human-like programs or machines. Instead, under today's technology, we are much more likely to encounter bots and AI in the form of personal digital assistants (like Apple's Siri) or algorithms which dynamically predict customer behavior (like those used by Netflix, Amazon or Facebook). Even the most cutting-edge technologies, like DeepMind's AlphaGo, IBM's Watson or Tesla's self-driving cars, are relatively primitive in comparison to the idea of an artificial, human-like brain. But leaving these limitations to the side, there is no reason why an AI (regardless of its sophistication) would necessarily pose such a dramatic challenge to the existing legal regime. Whether processing is performed by a human or by an AI, there is, in practice, often no significant difference to either the processing activities or their results. Therefore, even if a science-fiction-style AI would exist one day, we should not immediately assume that its actions would need to be regulated separately from processing which uses conventional methods. Such an assumption would risk creating a complicated and unhelpful distinction in data protection law, actively harming the development of socially-desirable technology and processing methods. To place the problems in terms of cliché, we must avoid both reinventing the wheel and throwing the baby out with the bathwater.

To this end, it is important to consider how the GDPR[1] regulates bots and AI. The recitals explicitly state that, to avoid circumvention, the law should be "technologically neutral and should not depend on the techniques used."[2] This philosophy underlies the majority of the GDPR: general rules which cover all forms of processing. The difficulty with this approach is that, while not always the case, there are some scenarios where processing by bots and AI could produce substantially different results to processing performed by conventional methods. In such a situation, general rules may seem inappropriate, e.g., because they are too burdensome or because they do not provide an adequate protection for data subjects. It appears that the drafters of the GDPR had a similar concern. Despite the general philosophy, certain provisions in the law are directly aimed at dealing with certain specific problems. The provisions most relevant to this chapter are usually phrased in terms of

[1] Regulation (EU) 2016/679 of the European Parliament and of the Council of 27 April 2016 on the protection of natural persons with regard to the processing of personal data and the free movement of such data, and repealing Directive 95/46/EC (General Data Protection Regulation) [2016] OJ L119/1.

[2] GDPR, Recital 15.

"automated processing" or in reference to "profiling." In addition to these specific provisions, the GDPR, as part of the risk-based approach, places an obligation on controllers to evaluate their processing activities. These provisions are particularly interesting as they form a co-regulatory approach; they do not specify what rules must be imposed, but instead demand that the controller designs their own, based on the factual analysis.

These three different approaches must be further examined before we can draw any firm conclusions about how best to regulate bots and AI. Section 2 of this chapter will begin by framing the technological landscape of the debate. It will argue that there is no clear definition of the terms "AI" or "bots" and examine how processing by such programs may differ from (or be similar to) processing by conventional means. Section 3 of this chapter will explore how this technological framework is governed by the legal framework. As noted above, the GDPR is, primarily, a technologically-agnostic approach to regulation. In principle, it regulates processing by bots and AI in the same was as processing by conventional means. In addition to this, Sect. 3 will also examine the other approaches in the GDPR. These will be identified as "the specific regulatory approach" (i.e., rules that are intended to regulate AI and bots in particular) and the "co-regulatory approach" (i.e., rules that delegate the responsibility of identifying risks and creating processing to controllers). This chapter will look at how these approaches are implemented in the GDPR and how that regulation affects bots and AI. Finally, Sect. 4 will contrast the approaches to provide a wider analysis and consider how they can best be utilized and balanced to protect both data subject rights and the interests of controllers and processors who wish to use bots and AI.

2 The Technological Landscape

The first hurdle that must be addressed in any discussion involving bots and AI is terminological. The problem is that there is almost no academic agreement on how the terms "bots" and "AI" should be defined. Some of the debates are semantic, whereas others are technical in nature. This chapter will not attempt an exhaustive review of the different definitions. Instead, this Section will examine the general trends of these definitions and attempt to propose some form of unified terminology that can be used to debate the topic.

The broadest definitions of AI require programs which can simply "perform tasks that would require intelligence if done by people," including "common sense tasks" (e.g., speaking human language or making sense of a novel's plot) or "expert tasks" (e.g., performing a medical diagnosis or building a computer program).[3] Alternative definitions require a computer to think in a logical or "human" way—or simply be able to act as though it does.[4] These definitions cover a wide range of possible pro-

[3]Sartor (1993), p. 15.

[4]Russell and Norvig (2016), p. 2.

grams. For example, it would clearly include the perfectly human-like AIs of science fiction. In the real world, however, there may be more ambiguity. There are clearly some programs which are capable of fulfilling the broad definitions. For example, digital assistants (including Apple's Siri, Amazon's Alexa and Microsoft's Cortana) are capable of at least limited understanding and speaking of human language, a "common sense task." However, there is also some ambiguity as to how well these tasks need to be completed. For example, Botnik Studios creates programs that analyze text and suggest new sentences based on identified patterns. In this way, Botnik's programs are actively writing text in the style of certain authors. Should this qualify as an expert task? Even if it does, Botnik publish collections of the most ridiculous sentences that are produced by their programs.[5] While many of the sentences are perfectly sensible, many are not and the programs would be extremely unlikely to pass the Turing Test. Should the programs which are used by Botnik be considered an AI, if only a weak or limited one? If we were instead to use a definition which refers to "humanity" in some way, we are also faced with certain philosophical questions. How does one (or, indeed, how could we) define "thinking in a human way"?

Meanwhile, it is even harder to find a fixed definition for the term "bot." Common usage covers a wide spectrum of products, including bots designed to send out mass numbers of emails ("mailbots"), bots designed to do housework (e.g., robot vacuum cleaners) and even bots that are designed to produce a shortened version of a news article.[6] As with the term "AI," there does not appear to be any universally-agreed-upon definition within the computer science or programming communities. Indeed, many of the programs that are described as bots may well fit some of the broader classifications for AI identified above (e.g., the AutoTLDR bot, which is programmed to identify key passages from news stories, a common-sense task). There are also dictionary definitions of the term "bot," but these are often contradictory and it is questionable how useful these are in an academic setting.

Nwana (who herself does not use the term "bot," but instead refers to "software agents") argues that there is no chance of reaching a consensus on the definition.[7] This argument is persuasive and applies equally to AI. Finding a clear line which would allow us to describe a program as either a "bot" or an "AI" is unlikely to be possible. However, we can make some general observations. First, based on the common usage and understanding of the terms, it is possible to draw a number of characteristics which, when taken together, can indicate that a program should be considered a bot or an AI. For a bot, these characteristics could include whether the program is self-executing, whether it performs bulk tasks, whether it deals with Big Data, whether it acts without interference from a human, or whether it operates on a wider network. For AI, these characteristics could include whether the program involves some form of machine learning or heuristics, whether the program performs

---

[5] See, e.g., Botnik's chapter written in the style of Harry Potter. Available at: http://botnik.org/content/harry-potter.html. Accessed 3 January 2018.

[6] See, e.g., the Reddit bot AutoTLDR. Available at: https://www.reddit.com/r/autotldr/comments/31b9fm/faq_autotldr_bot/?st=j041su3v&sh=295425b7. Accessed 3 January 2018.

[7] Nwana (1996), Part 4.

common-sense or expert tasks, or whether the program attempts to portray itself as "human." This approach does not require a program to meet every characteristic in order to be classified as a bot. Rather, it allows us to look at a program and consider whether, all things considered, such a description would be appropriate.

It is also proposed that, in general, the term "bot" should be preferred to the term "AI." There is clearly some overlap in the terms, but it is likely that most programs that exist in the current state of technology will fit into the term "bot," while many fewer will fit into "AI." This is helpful as there may often be a lot of overlap where a program could be considered both an advanced bot and a primitive AI. Use of the term "bot" therefore allows us to hold a more effective regulatory discussion. Further, the term "AI" carries unhelpful connotations. It is difficult to discuss regulatory limits and compromises if those discussing the problem are thinking about personal digital assistants with the intelligence of Iron Man's Jarvis rather than Apple's Siri. Finally, the term "bot" is somewhat less philosophical than some interpretations of the term "AI." While everyone can agree whether or not a program operates without human interference, whether or not something can think in a way that should be described as "human" is much harder to pinpoint in a regulatory manner.

These conclusions still present a problem for any regulatory discussion. To implement the definitions given above into law would create significant legal uncertainty. Meanwhile, creating a more concrete definition for the purposes of legal regulation would risk creating an artificial hardline distinction. Given the potential for overlap, the definitions above could be seen as providing a gradient of intelligence, with "normal" computer programs being the least intelligent and a true, science-fiction AI as the smartest. At what point do we deem the program "smart enough" to warrant independent regulation?

These definitions are, however, sufficient for us to examine how processing by bots and AI may differ from conventional processing methods. And, indeed, as discussed above, it is not always obvious that any differences may actually exist. As part of the Future Regulation of the Industrial Internet (FRII) project, we ran a serious of workshops analyzing the impact of autonomous shipping on data protection. These workshops gathered experts in both technology and the shipping industry and the results of their discussions was clear. Although autonomous cargo ships would process personal data, they would not do so in a way that is meaningfully different to the way that crewed cargo ships process the data.[8] This is particularly interesting as the bots and AI which are integral to autonomous shipping may be incredibly sophisticated—they will need to deal with large numbers of variables to sail safely, may process information in a dynamic fashion, will run multiple systems and will require advanced and responsive cyber-security measures. Equally, it is easy to imagine a fairly unsophisticated bot or AI which processes personal data (even in bulk) in a way that is not novel. For example, a bot which records when a worker starts and stops their shift and uses this to calculate a worker's monthly salary would be no

[8]This conclusion was reached at the FRII Analysis Workshop held at Aalto University, Helsinki, Finland on 14 June 2017 and reinforced at the FRII Seminar held at the Hanken School of Economics, Helsinki, Finland on 11 September 2017.

different to a human recording this information on a time card and making the same calculation.

On the other end of the scale, however, we can see situations where bots and AI may dramatically revolutionize the processing of personal data. Unlike conventional processing, which is limited by human capabilities, bots and AI are capable of analyzing vast amounts of information. Already, the algorithms behind Big Data bots can find patterns and connections that would be practically impossible for human analysts to identify. Additionally, while areas such as autonomous shipping involve very predictable uses of personal data, other processing may be more unpredictable, particularly if the processing is novel or involves dynamic scenarios. While a bot or an AI performing such a processing may come to the same conclusion as a human, this will vary depending on the situation.

The impact of processing by bots and AI, then, must be evaluated on a case-by-case basis. Notably, this does not necessarily depend on the sophistication of the bot or the AI in question. Rather, the question is as to the actual processing in question. It is with this in mind that we must turn to examine how the GDPR regulates processing by bots and AI.

3 A Look at the GDPR

This Section will break the GDPR into three main approaches: approaches which cover all forms of processing ("the general approach"), approaches which are specifically aimed at processing by bots and AI ("the specific approach") and approaches which are designed to share the regulatory burden between the legislature and private actors ("the co-regulatory approach"). It will look at different provisions which demonstrate these approaches and consider how they may be applied to processing by bots and AI.

One of the stated purposes of the GDPR is to "ensure a consistent and high level of protection of natural rights and to remove the obstacles to flows of personal data within the Union."[9] The GDPR further states that the processing of personal data "should be designed to serve mankind."[10] We must measure the effects of the different approaches against these goals. It must also be remembered that while serving mankind means that bots and AI should not be used to prejudice people's rights, part of this concept also includes ensuring that data protection law does not unduly prevent the freedom to conduct a business. While the law should protect data subjects from the potentially harmful impacts of bots and AI, it must also encourage and permit technology to grow in a socially desirable way. We must examine how each regulatory approach is used to attempt this balancing act and attempt to identify any reasons why they may fail to adequately do so. To this end, this Section will look

[9]The GDPR, Recital 10.

[10]The GDPR, Recital 4.

at different examples of the regulatory approaches, as embodied by the GDPR, and attempt to draw some conclusions about how these approaches function.

One observation that could be made is that the dichotomy between general regulations and specific regulations could be seen as a parallel to the rules-based regulation vs. standards-based regulation debate.[11] However, these discussions are distinct. For example, there are many examples of general regulations which are rules based (e.g., the rules relating to the rights to information in Articles 13–14, discussed below). The focus in this chapter is on the general, processing-method-agnostic regulatory approach as opposed to the specific, bots-and-AI-focused regulatory approach and the co-regulatory approach.

3.1 The General Regulatory Approach

As has already been stated, the majority of the provisions in the GDPR apply to all processing of personal data, regardless of how that processing takes place. For example, the definition of "personal data" in Article 4 (1) refers to "any" information, while the definition of "processing" in Article 4 (2) refers to "any operation...whether or not by automatic means." Both of these definitions include lists of examples, but they are not exhaustive. This form of language can be seen throughout the GDPR and clearly shows the principle of general applicability. Clearly it will not be possible to examine every provision in this chapter, but this sub-section will attempt to survey a representative sample of the general provisions and the impact that they may have.

There are situations where these general rules will apply smoothly to processing by bots and AI. The immediate starting point is that if the introduction of AI and bots does not significantly alter the nature or result of the processing, the general rules will apply in more-or-less the same fashion as in a conventional processing scenario. For example, a bot which reviews student attendance by collating registers and flagging any student who has been absent for more than a certain percentage of days will not operate any differently to a teacher collecting the results together. Equally, as noted above, the use of autonomous ships will not cause any special difficulties under the general rules as the introduction of the bot will have little impact on how the data is actually processed. When such bots are used, therefore, the question will not be "How do the general rules govern bots and AI?" but simply "Are the general rules correctly drawn?"

What we must consider are the alternative cases, where the processing by bots and AI creates a significantly different result to processing by conventional methods. There are some situations where this occurs, but where the general rule still provides a satisfactory result. A notable example of this is Article 5 (1), the principle of lawfulness, fairness and transparency. As noted by Recital 39, "any processing of personal data should be lawful and fair." It seems hard to argue that we would not want to apply this principle to the processing of personal data by bots and AI. This

[11] See, e.g., Maxwell (2015), pp. 212 et seq.

argument is not limited to provisions which simply set out general principles. For example, Chap. 5 of the GDPR sets out many comprehensive rules about the transfer of personal data outside of the EU. As stated in Article 44, these rules must be applied to ensure that the protection of data subject rights is not "undermined." This reasoning is not dependent on the processing method. For example, the factors which should be considered when issuing an adequacy decision under Article 45 should not depend on the amount of data being transferred by single controllers, or whether data will be transferred using a bot sending an email or a human sending a letter—either the third country offers sufficient protection for personal data or it does not.

We should conclude that it is possible for both principle-based provisions and rules-based provisions to be generally applied in a desirable way. Some provisions, however, are more controversial. The definition of personal data under Article 4 (1) is very wide and is deliberately written to cover all personal information, regardless of the where and how of the processing. This definition is further expanded by decision of *Breyer*.[12] In this case, the Court of Justice of the European Union ("the CJEU") said that it was not necessary for a data subject to be identifiable by the information alone, nor that all information enabling identification be held by a single person.[13] Rather, information is personal unless the chances of identification are "insignificant" because gathering the necessary extra information would either be illegal or would require such a "disproportionate effort in terms of time, cost and man-power" as to be "practically impossible."[14] Interestingly, the test itself seems to assume a conventional processing environment—by placing a focus on "time, cost and man-power," *Breyer* neglects to consider the idea of a bot which can effortlessly work in the background to find the necessary information. Nevertheless, this is a general rule which will apply to all processing activities. Under *Breyer*, processing in a conventional scenario will be more likely to fall outside of the GDPR than processing which involves bots and AI. It will, for example, likely be considered disproportionate for a human to comb through Big Data records to find all of the information that could be used to identify the data subject. However, data-mining bots are designed to crawl through such data sources, or even through the Internet itself, and gather different pieces of information with minimal effort and cost.[15] The introduction of these bots into the processing environment means that the test introduced in *Breyer* is a relatively low one.

There are legitimate arguments to be made to support such a low barrier for processing in a bot and AI environment. Since bots have such a powerful capability for combining information, one can argue that we should be ready to apply the protections in data protection law even at such a relatively low stage. This argument is not, however, universally accepted. One potential counterargument is that, given

---

[12]C-582/14 *Breyer v Bundesrepublik Deutschland* (Second Chamber, 19 October 2016). Published in the electronic Report of Cases.

[13]C-582/14 *Breyer v Bundesrepublik Deutschland*, paras. 41–43.

[14]C-582/14 *Breyer v Bundesrepublik Deutschland*, para. 46.

[15]See, e.g., Googlebot. Available at: https://support.google.com/webmasters/answer/182072?hl=en. Accessed 26 November 2017.

the many burdensome obligations contained in the GDPR (and the strong penalties for failing to meet with the required standards), it is impractical to have such a low barrier for personal data. This is especially relevant where controllers may be unaware that the information is personal data. This could occur, for example, if they are unaware that the use of a Big Data bot could lead to identification. Given that it is possible to use Big Data bots to identify individuals from relatively innocuous pieces of information, the implications of *Breyer* may mean most data has the potential to be considered personal.

Further, it could be argued that the introduction of bots and AI has the potential to break this interpretation in many cases. The court did not say whether the processing activities are "likely" to identify the data subject, but rather whether the odds of identification are "insignificant." One could argue that, since the majority of controllers have at least the possibility of introducing data-mining bots into their operation if they so choose, the odds may never be truly "insignificant." Indeed, given that Big Data analyses often result in unexpected or previously unpredictable linkages (even where companies believed their data sets to be completely anonymous),[16] one could argue that the only way to know whether the odds of identification are insignificant is to actually try. Ironically, this would actually result in a greater intrusion for data subjects, with potential controllers attempting to identify data subjects just to see if it is possible. Any such interpretation of *Breyer* would be incredibly difficult to work in practice, would place incredibly harsh burdens on actors who would not normally consider themselves to be controllers and would likely have a significant impact on the development and commercial viability of bots and AI.

This highlights the difficulty with attempting to strike the correct balance with general provisions. It is difficult to come up with a single rule (especially one which produces a binary outcome) which can adequately regulate scenarios where identification is almost impossible and scenarios where identification is almost certain (if attempted). Whether we wish to implement stronger or weaker protections on either side, it is clear that both will have different relevant policy considerations. Given the potential impact of treating information as personal data (both in terms of the protection due to the data subject and the obligations to be placed on the potential controllers and processors), it seems simplistic to merely ask whether the possibility of identification is "insignificant", regardless of the scenario.

The use of bots and AI may also unbalance the administrative compromises drawn up by general provisions. For example, Articles 13–14 set out certain rights to information. These Articles include lists of details which must be provided to data subjects, depending on how the data was obtained. In addition, Article 15 sets out two broad rights to access to information, including information about how the processing takes place and a right to obtain a copy of the personal data that is being processed. The policy justifications for these rights apply just as clearly to information processed by bots and AI as to information processed in a conventional scenario. Further, as bots and AI are often designed to discover new information about data subjects, these rights may be even more important in such an environment as they will allow data

[16]See, e.g., Mayer-Schönberger and Cuckier (2013), pp. 154 et seq.

subjects to hold controllers accountable. However, the use of bots and AI can allow very small controllers to process the personal data relating to enormous numbers of data subjects. If even a small percentage of these data subjects make requests to controllers, those controllers may quickly find themselves overwhelmed. While the general principle of access to information is therefore still desirable, we may wish to strike a different balance or impose different specific requirements, depending on the nature of the processing activities. This difficulty could also apply to Article 17 (the right to erasure/"right to be forgotten"), where the Council of Europe had extensive debates on how best to balance the protection of data subjects and the burden imposed on controllers.[17]

The risks in relation to these particular Articles should not be overstated. It is, for example, also possible to use bots to minimize the burden created by the information rights by automating a response. Nevertheless, by imposing a general rule, it becomes much more difficult to properly balance the level of protection necessary and the administrative burden on the controller.

Another general provision worth noting is the concept of consent. Much has been said about the difficulty of consent and Big Data. For example, it has been argued that a model based on consent is not well-suited to the information world,[18] or that Big Data's ability to discover unpredictable links means that it is impossible to obtain sufficiently informed consent.[19] As many (arguably most) Big Data programs should be considered as bots, these debates will be equally applicable here. Indeed, many issues that arise with Big Data bots are likely to be seen in other types of bots and AI as well. Where a program independently modifies the algorithm which it uses to process personal data, even if it does not utilize Big Data, it could be argued that it will be impossible to obtain truly informed consent. Even if the data subject were aware of the ways in which the algorithm could change, these changes may be so substantial that the original consent cannot cover the processing activity. If this were to occur with a human processor, that human could simply know that they were required to obtain a new consent; such self-reflection may not be possible with a bot or an AI.

Under these interpretations, the law poses a real challenge to the use of bots and AI. While other justifications exist for the processing of personal data, consent is often viewed as having a notably prominent position.[20] If many bots and AI are fundamentally incompatible with the legal definition of consent, this will have a significant impact on the viability of research and development into those technologies. One could argue that this is actually desirable. As with the rules relating to International Data Transfers or the principle of lawfulness, fairness and transparency, one may argue that we should not expect to treat processing by bots and AI any differently. The rules for consent are drawn in a way that provides a certain level of protection to data subjects; controllers should not be able to avoid this by using bots

[17]See, e.g., Council of the European Union (2014).

[18]See, e.g., Pearce (2015).

[19]See, e.g., Rubinstein (2013).

[20]See, e.g., Míšek (2015), or Article 29 Working Party (2011), p. 7.

or AI to perform processing activities. If we cannot create bots and AI which meet these criteria, we should not be developing them in the first place. While this may be true, it precludes the possibility of finding a third option—one which provides an equal level of protection, but which does so while also allowing for the potential peculiarities of processing by bots and AI. Assuming that such a test could be agreed upon, there may be benefits to creating a specific regulation to deal with consent in a bot and AI environment.

Having accepted that bots may encounter some difficulties with the concept of consent, it must also be noted that bots will make certain aspects of consent easier. For example, the GDPR, Article 7 (3) states that data subjects have the right to withdraw consent at any time and that it "shall be as easy to withdraw as to give consent." Pearce argues that increasing complexity online makes it harder for data subjects to keep track of their various consents,[21] but this issue can be addressed by the use of bots. For example, bots can be created as consent management tools. It is well within the scope of technology to create a bot which will respond to data subject requests to withdraw consent and then automatically delete the relevant data.[22] Equally, there are broader consent management systems being designed that are intended to help data subjects manage their various consents[23] which will likely utilize bots to at least some extent.

The use of the general regulatory approach in the GDPR, then, produces a mixed result. In some situations, the use of bots and AI will not make any difference to the provisions; in others, the provisions are even more valuable when it comes to innovative bots and AI; while in others, the general provisions may make it more difficult to find an appropriate balance in a given factual scenario. Equally, we must recognize that the use of bots and AI will, in some cases, make it easier to comply with these general provisions. We should, therefore, conclude at this stage that it may often be appropriate to use general provisions to regulate processing by bots and AI, even when that processing is radically different to processing in a conventional environment. Nevertheless, it is important to consider whether a specific regulatory tool may be more appropriate, or may offer more room for certain policy considerations.

3.2 *The Specific Regulatory Approach*

Having examined the general regulatory approach, we must now turn to those areas of the GDPR that use the specific regulatory approach. Under this approach, a provision is written to deal with specific scenarios or issues and will only apply to certain

[21] Pearce (2015), p. 151.

[22] One issue with such a bot is that it will require appropriate data security measures to, e.g., ensure that the removal of consent is actually coming from the relevant data subject. The depth of this security will depend on the nature of the data and the processing involved, but should be manageable in most scenarios.

[23] See, e.g., the MyData project. Available at: https://www.lvm.fi/documents/20181/859937/MyData-nordic-model/. Accessed 29 November 2017.

types of processing. The GDPR does not include any regulations which are explicitly narrowed to processing by bots and AI. It does, however, include a number of provisions that target "automated processing." As will be argued below, we can find that many provisions are actually intended to address the issues related to bots and AI. In this way, and given the difficulties with outlining a legal definition of bots and AI described above, the provisions can be seen as a form of "indirect" specific regulation governing processing by bots and AI.

One provision which is closely aligned with processing by bots and AI is Article 22, i.e., the right relating to automated individual decision-making. This Article states that data subjects have a right "not to be subject to a decision based solely on automated processing, including profiling, which produces legal effects concerning [them] or similarly significantly affects [them]." As one of the common characteristics of a bot is the ability to work autonomously, without human intervention or interference, this Article immediately invokes the idea of bots and AI. This could very easily include, for example, bots which are designed to take an insurance application form, scan it for the relevant information and then give an insurance premium quote.

Article 22 is notable because it would not be possible, or desirable, to impose a general restriction against individual decision-making. Such an abstracted rule would, for example, mean that any sufficiently-important decision made by a human being would be unlawful unless the decision were checked by a third party or if one of the exceptions in Article 22 (2) applied. It would be incredibly hard to argue that such a rule was desirable, or even conceptually possible to enforce. Because we are only concerned with decisions made automatically (i.e., by bots and AI operating without the supervision of a human), this problem can only be dealt with by specific regulation.

Disappointingly, there are a number of considerable ambiguities in Article 22.[24] For example, it does not explicitly define when a decision is based "solely" on automated processing. This is an unfortunate omission as the definition could have clear policy implications. For example, a strict interpretation of the word "solely" could be that Article 22 will apply unless a human actor has carefully evaluated the entire decision, including the inputs, the logic and the outcome. Under such a definition, the use of bots and AI to make certain types of decisions will be significantly less attractive, but there will be a much stronger protection for data subject rights. By contrast, a more relaxed interpretation of the word "solely" could be that Article 22 will apply provided that a human being has some sort of oversight over the decision-making process. One possible interpretation could be that the decision is not made "solely" as long as a human looks at the output and then approves it as being correct, without actually having to analyze the input. This wider interpretation would provide a weaker protection for data subjects, but would provide much more viability for the development and adoption of bots and AI.

[24]For a wider discussion of this Article, see Kamarinou et al. (2016).

This issue has been addressed by the Guidelines on Article 22 issued by the Article 29 Working Party.[25] The Guidelines say that a decision is made "solely" if it lacks any meaningful human oversight. To avoid this, the original version of the Guidelines stated that oversight must be carried out by an actor who had the "authority and competence to change the decision" and that the actor should "consider all the available input and output data." They further stated that the decision would not be made solely by the automated process if the bot produced something which was "in effect a recommendation concerning a data subject" and that the final decision is made based on a review of that recommendation while taking into account other factors.[26]

By requiring the human actor to review all of the available input and output data, this interpretation takes a very data-subject-friendly approach. There are clear reasons supporting this approach. The protection of personal data is a human right guaranteed in the EU Charter.[27] It must also be remembered that Article 22 is limited to decisions which produce "legal" or "similarly significant" effects. This requirement will narrow the Article's application, even if the term "solely" is given a wide meaning. The required impact of the effects is uncertain—the GDPR does not provide any definitions and the Article 29 Working Party has noted that there is some ambiguity in the term, although it proposes that the impact must be "sufficiently great or important to be worthy of attention."[28] Regardless of the exact definition, it is clear that Article 22 is only intended to deal with situations where there is a high risk of prejudice to data subject rights and so should be interpreted in a way which provides a strong protection.

Despite this narrowing of scope, Article 22 will still apply to a wide variety of situations. Recital 71 gives the example of the automatic refusal of an online credit application but, as the Article 29 Working Party pointed out in the original version of the Guidelines, this could potentially include anything from renting a city bike while on holiday to obtaining a mortgage.[29] If such a wide definition of Article 22 is accepted, this may significantly harm the adoption and development of bots. It is notable that, in their revised version of the Guidelines, the Article 29 Working Party has arguably relaxed their interpretation of the word "solely", now simply requiring that a human "consider all the relevant data".[30] This change makes sense; if a human must give a full review of the inputs and outputs for every decision, there would be no commercially viable reason to utilize a bot or AI in any sufficiently serious decisions unless one of the exceptions in Article 22 (2) applies. Where this is paired with a wide interpretation of whether something similarly significantly affects a data subject, this will mean that there are large areas of processing where the use of bots and AI are entirely undesirable.

---

[25] Article 29 Working Party (2017a), as revised by Article 29 Working Party (2018).

[26] Article 29 Working Party (2017a), p. 10.

[27] Charter of Fundamental Rights of the European Union, Article 8.

[28] Article 29 Working Party (2018), p. 21.

[29] Article 29 Working Party (2017a), p. 11.

[30] See Footnote 28.

Article 22 (2) contains three exceptions, namely that the processing is necessary for entering into, or performing, a contract; that the use of automated decision-making is authorized by EU or Member State law and subject to suitable safeguards; and that the data subject has explicitly consented to the decision. The first two of these exceptions are extremely narrow. As a result, an overly wide interpretation of Article 22 (1) will likely result in a large reliance on consent. Given the issues relating to consent discussed above, this will clearly cause issues. It has been said that "I have read and agree to the Terms" is the biggest lie on the Internet.[31] By creating rules which are so restrictive that controllers always ask for consent (which will very frequently be given without consideration of the impacts), there is a risk that any protection offered by Article 22 will be rendered inefficient. In this way, attempting to increase the level of protection will actually reduce it.

It would also be very easy to draw the line too narrowly to provide an efficient protection. The Commission has previously noted that human decision-makers are at risk of attaching "too much weight" to a decision made by a bot, even if they would have been critical of the decision if it had been made by another human.[32] As noted by Bygrave, it cannot be accepted that a decision is not made solely by a bot if a human has merely rubber stamped the result without actually assessing it.[33] The real solution should, therefore, prefer a middle-ground approach. One possible solution may include oversight of the decision process and a reasonableness check of the outcome, but without requiring that a human essentially remakes the original decision.[34] It is likely that this Article and its interpretation will be the subject of considerable debate and litigation.

A related provision is Article 15 (h). Where a processing operation would fall under Article 22, the data subject is entitled to obtain information about the existence of the automated decision-making, details about the significance and envisaged consequences of the processing, and "meaningful information about the logic involved." This provision is also relatively ambiguous. The Article 29 Working Party states that controllers should "find simple ways" of telling the data subject about the rationale behind or criteria involved in making the decision, without necessarily giving a full and complicated explanation of the algorithm. However, it also notes that "complexity is no excuse for failing to provide information" and concludes that the information should "be meaningful to the data subject."[35]

Although data subjects are entitled to significant amounts of information, it is notable that there is no right to know the logic behind a decision made by a human. As with Article 22 itself, then, Article 15 (h) appears to exist to counter the perceived

---

[31] See, e.g., Terms of Service: Didn't Read. Available at: https://tosdr.org/. Accessed 3 January 2018; biggestlie.com. Available at: http://biggestlie.com/. Accessed 3 January 2018; and Obar and Oeldorf-Hirsch (2016).

[32] European Commission (1992), p. 26.

[33] Bygrave (2001), p. 26.

[34] Such a test could draw inspiration for the rules of judicial review in England and Wales: See *Associated Provincial Picture Houses v Wednesbury Corporation* [1948] 1 KB 223 and subsequent case law for more details.

[35] Article 29 Working Party (2018), p. 25.

extra dangers to data subject rights when decisions are made by bots and AI. Further, as with Article 22, it would arguably be extremely undesirable to implement such an obligation on human decision making. Although transparency in relation to decisions is generally a good principle, introducing a general legal obligation to provide an explanation of the logic behind decisions made by humans would be very onerous and difficult to fulfill. Even where those decisions have the required significant impact on the data subject, these decisions may justifiably be made on relatively subjective or emotional grounds that could not be explained (e.g., deciding to hire one of two equally-capable and qualified candidates for a job because one fits into the work environment well, while the other does not).

Based on the above analysis, we can conclude that the core concern being addressed Articles 15 (h) and 22 is that processing by bots and AI introduces additional risks which must be specifically regulated. To deal with this issue, these provisions introduce rules and safeguards intended to mitigate these risks where the decision is sufficiently important. Although there is room to debate the balance attempted by the provision and how best to implement it, it is clear that these concerns could not be addressed through a general regulatory approach.

Another concern that is addressed through the specific regulatory approach, at least to some extent, is profiling. Under Article 4 (4), profiling is any "automated processing" which uses personal data "to evaluate certain personal aspects" of a person. This applies "in particular" to analyses of work performance, economic situations, health or certain habits and interests. Unlike Article 22, there is no explicit requirement that the profile be created "solely" by automatic processing. Nevertheless, the reference to automated processing in Article 4 (4) immediately invokes the idea of a program which combines information (possibly from a Big Data source), runs it through an algorithm and creates a profile for the data subject. While this program will not necessarily be a bot, there will be a significant overlap. One key area is advertising; profiling is (in)famously used to provide personalized adverts to individuals online. These systems are designed to operate for millions of customers and the only way to run such a system in an economically viable fashion would be to automate it with bots. Alternatively, profiling systems that are designed to dynamically update will likely be managed by bots, as will those that perform real-time evaluations based on Big Data analytics.

What is notable about the inclusion of profiling in the GDPR, however, is that while "profiling" is specifically identified, it is almost always only used as an example. In the substantive Articles of the GDPR, the word "profiling" appears 11 times. Of those 11 times, one is a definition and one mandates that the European Data Protection Board should issue guidelines, recommendations and best practices for "decisions based on profiling pursuant to Article 22 (2)." The remaining 9 references (one of which is a title) merely use profiling as an example of processing. For example, Article 22 refers to decisions based on automated profiling "including profiling," while Article 21 (2) gives a right to object to processing used for direct marketing purposes, "which includes profiling to the extent that it is related to such direct marketing."

At first glance, this feels redundant. By definition, "profiling" is a form of processing. Therefore, a data subject's right to object to processing must include the ability to object to profiling. Moreover, as the lists are not exclusive, the provisions which specifically refer to profiling, are often, in essence, actually general rules. We must, then, ask why the specific regulatory pointer was included when it does not provide any modification or extra-legal protection. To answer this question, it is helpful to examine the legislative background of the GDPR. For example, in the European Parliament's report on the proposed GDPR, the Committee on the Internal Market and Consumer Protection stated that profiling "should, in principle, only be permitted where there is a solid legal basis" and "stresses" that data subjects should have the right to information and erasure, "especially for profiles compiled for commercial purposes."[36] The European Parliament noted that profiling was "a major trend in the digital world."[37] Meanwhile, the Commission's first draft of the GDPR[38] made reference to the Council of Europe's Recommendation on profiling.[39] This Recommendation pointed out that profiling often makes it possible to "generate new personal data" on the data subject in a way that can be invisible.[40]

The pointers to profiling, then, could be seen as simply being a way for the legislature to emphasize their concern over certain issues, without necessarily wishing to create different rules for them. This approach can be seen elsewhere in the GDPR. For example, "personal data" under Article 4 (1) states that personal data is any information relating to an identified or identifiable natural person, "such as" a number of examples. These items, as with the references to profiling, are simply examples which were seen as being of particular importance or significance during the legislative process. This approach has certain advantages in that it can draw attention to particular cases, but (as noted above) does not actually provide any actual extra-legal protection or produce any legal effects that are unique to bots and AI.

The specific regulatory approach therefore can be used in a number of ways, from providing specific rules for specific scenarios to simply highlighting the importance of certain issues without actually providing different rules. In theory, the former allows for a balanced discussion of the specific issues surrounding bots and AI. Ideally, these discussions would happen at a legislative level, with good regulation providing sufficiently clear and understandable rules to support the rule of law. Even where these provisions are ambiguous, however, they allow for a more nuanced discussion of the specific issues by courts or other bodies, although this may weaken the democratic legitimacy of the final interpretation. A second use of regulatory provisions has a subtler impact; it does not directly affect the actual content of the law, but may act as a particular emphasis, drawing the attention of controllers or enforcement agencies.

---

[36]European Parliament (2011), p. 21.

[37]European Parliament (2011), p. 9.

[38]European Commission (2012), p. 9.

[39]Council of the European Union (2010).

[40]Council of the European Union (2010), p. 6.

3.3 *The Co-regulatory Approach*

The third approach identified in this chapter is the co-regulatory approach. Unlike the general or specific regulatory approach, this technique does not actually lay out firm legal rules. Rather, co-regulation has been described as "various kinds of collaboration between state and private actors in some aspect of the regulatory process, where there is at least some form of legal enforceability."[41] These provisions have also been described as "meta regulation"[42]—i.e., "regulation...that regulates another form of regulation," in this case "legal meta-regulation of internal corporate self-regulation."[43] Of the approaches discussed in this chapter, this is the most flexible approach as it, in effect, creates an obligation for private actors to invent "do-it-yourself" legislation. This flexibility, however, also comes with certain risks—by necessity, the co-regulatory approach is quite vague as to requirements and there is a risk that controllers will fail to provide adequate rules and safeguards.

One important co-regulatory provision is Article 25 (1), which introduces the principle of Data Protection by Design. This states that controllers have a duty to implement "appropriate technical and organizational measures" to ensure compliance with the GDPR, protect data subjects and integrate data protection principles, both when actually performing the processing and when planning it. Importantly, the scope of this obligation scales to the scenario. A controller is obliged to take into account "the state of the art, the cost of implementation and the nature, scope, context and purpose of processing," as well as the risks to data subjects. This Article derives from the wider principle of Privacy by Design, which sets out seven foundational principles, such as "proactive not reactive," "end-to-end security" and "visibility and transparency."[44]

The concept of Data Protection by Design may seem as though it should be so obvious that there is no need to legislate it. Given how onerous some of the obligations in the GDPR can be, one must keep these rules and principles in mind from the beginning or have no hope of actually complying with them. Nevertheless, by creating a legal obligation to comply with Data Protection by Design, the law has performed a number of roles. First, it has underlined the importance of these principles and increased the chances of awareness. Controllers who may previously have simply considered it a "good idea" to think about data protection at an early stage will have an extra encouragement to do so. It also encourages controllers who may have been reckless with data protection issues to actually think about the impacts of their processing. Secondly, it allows enforcement agencies to penalize controllers who have not only failed to adequately protect data subject rights, but who failed to place give those rights due weight and consideration.

Thirdly, and extremely importantly, Article 25 (1) means that the exact legal obligations placed on the controller depends entirely on the factual scenario. One of

[41] Binns (2017), p. 25.

[42] See Binns (2017).

[43] Parker (2007), pp. 14 and 98.

[44] See Information & Privacy Commissioner of Ontario (2013).

the conclusions from Section 2 was that it is difficult to generalize processing by bots and AI. By providing a framework for the controller's self-regulation, and creating the possibility of penalties for failing to adequately comply with that framework, it is possible to provide for this issue through legal mechanisms. For example, in the autonomous shipping example, Article 25 will not require particularly novel or dramatic self-regulation. By contrast, if the controller is attempting to create a bot or an AI which monitors customer behavior and builds profiles for each customer involving special categories of data to be used for marketing purposes, the law is capable of demanding much more rigorous actions from the controller.

Article 25 is complemented by Article 35, which sets out the rules for Data Protection Impact Assessments ("DPIAs"). Article 35 states that if processing "is likely to result in a high risk" to data subjects, the controller must perform an assessment which includes, inter alia, a "systematic description" of the processing, an assessment of the necessity of the processing, an assessment of the risks of the processing and the measures envisaged to minimize those risks. The law further sets out a number of situations where DPIAs are required and different factors that should be considered. The Article 29 Working Party has released guidelines on performing and evaluating DPIAs.[45]

As with Article 25, this provision creates a flexible regime that can be introduced when necessary. Although Article 35 states that DPIAs should especially be used in situations "using new technologies," the actual requirements and results of the DPIA will scale, depending on what the bot or AI in question is doing. The risk is that the flexibility in provisions like Articles 25 and 35 provisions will lead to inadequate protections. For example, while he accepts that DPIAs are useful instruments, Koops argues that the use of "open and fuzzy norms" and a lack of internalization of data protection principles by controllers means that Article 35 will simply result in a rubber-stamping procedure, rather than the creation of meaningful self-regulation.[46] This concern warrants serious consideration. It is not inconceivable that users of bots and AI may choose to merely perform lip service to the co-regulatory procedures. This issue is mitigated by some extent to the mandatory inclusion of a data protection officer in certain circumstances,[47] but these officers are not required in all situations where a DPIA may be required. Further, there is no reason to assume that a controller who would deliberately ignore the spirit of Articles 25 and 35 would appoint an officer who was likely to object to their doing so. The effectiveness of these provisions, then, may depend to some extent on enforcement—which will have to be seen after the GDPR has come into effect.

Not all co-regulatory provisions are subject to this criticism, however. Whereas Articles 25 and 35 place obligations on individual controllers to create internal regulatory regimes, Article 40 encourages the creation of codes of conduct for wider usage. Under this Article, "associations and other bodies representing categories of controllers or processors" should be encouraged by various Member State and EU

[45] Article 29 Working Party (2017b).

[46] Koops (2014), pp. 254–255.

[47] GDPR, Sect. 4.

bodies to draw up codes of conduct "for the purpose of specifying the application of [the GDPR]." Once created, a code of conduct can be formally approved by the Supervisory Authorities (if the code relates to only one Member State) or the Commission (if the code relates to multiple Member States). Approved codes are then collated in a register and made publicly available.

These codes may be very important for controllers who wish to use bots and AI. By adopting a code which explains how to implement the general regulatory provisions in a specific way, controllers will be able to benefit from increased legal certainty. Further, since the codes must be approved, they should be written in a way that provides a balanced and reliable protection for data subjects, rather than simply promoting the interests of the group who wrote the code. The GDPR offers multiple legal incentives for controllers and processors to comply with codes of conduct. For example, the use of an approved code of conduct can be used as a basis for a transfer of personal data outside of the EU under Article 46 (2) (e), or can be used to mitigate penalties for non-compliance under Article 83.

In these ways, codes of conduct are very similar to the certification method under Articles 42 and 43. This method allows for the creation and official approval of certification bodies which can certify controllers who comply with their schemes. Certification will have a number of advantages for controllers. First, the controllers will be able to advertise themselves as compliant with the GDPR, as demonstrated by the certificate. This will help to increase data subject confidence in their services. Secondly, the certification can (as with codes of conduct) be used to, inter alia, transfer data to non-EU countries and mitigate penalties. Finally, the certification process requires the controller to demonstrate that they comply with approved procedures, which increases the certainty that their actions are compliant with the law.

It is difficult to state exactly how these co-regulatory schemes will impact bots until we have seen the level of adoption. Certainly, there are a number of clear advantages. Because there is no need to draw a bright-line division between those to whom these rules must apply and those to whom these rules must not apply, it may be easier to create schemes which cover bots and AI despite the definition difficulties identified in Sect. 2. Further, as both codes of conduct and certification schemes are much easier to update than specific legislation, it is possible to create more precisely-detailed requirements, which can be balanced (and, importantly, relatively easily changed) according to the needs of the circumstances and situations. It would, for example, be possible to create one code of conduct for Big Data bots and another code of conduct for autonomous ships, and to update these codes of conduct as the technology improves or changes. However, there is a risk of over-saturation—if we create a code of conduct for every possible type of bot, it will create an overly complicated co-regulatory scheme. While the voluntary nature of compliance means that this does not cause as much difficulty as a complicated legislative regime, it still reduces the utility and desirability of these mechanisms.

There are doubts as to whether these systems will actually provide any real protection. Koops argues that, as of 2014, co-regulatory approaches were not sufficiently utilized, in part because regulators spent "relatively little effort to communicate best practices" and in part because it is uncertain if sufficient amounts of co-regulatory material were being produced.[48] It must be accepted that widespread adoption is necessary for co-regulatory methods like codes of conduct and certification methods to be effective. However, there are many examples of such schemes being widely accepted. For example, the Safe Harbor and Privacy Shield mechanisms (despite their flaws) have been widely adopted by US companies who wish to transfer data to and from the EU.[49] Equally, there are many widely-adopted standards in other fields, many of which are either directly or indirectly related to bots, AI and data protection.[50] Finally, as noted above, the GDPR introduces significant advantages to controllers who adopt these approaches, which should encourage their use.

In conclusion, it is hard to accurately predict how the co-regulatory approaches in the GDPR will affect processing by bots and AI. By offering such flexibility, this approach has the potential to be an invaluable part of the regulatory toolbox. No other approach identified in this chapter has the same potential to provide both clear guidance and flexible scaling. The real challenge will be ensuring that these approaches are, first, sufficiently widely adopted and, secondly, used correctly. Where there is a possibility for a useful and balanced code of conduct to be created, there is also the potential for an unhelpful and harmful code of conduct to be created; where there is a chance for a DPIA to provide a thoughtful and thorough consideration of the risks associated with a bot's processing of personal data, there is a chance for a DPIA to serve as nothing more than a rubber stamp.

4 Analysis

The discussion in Sect. 3 has highlighted a number of the strengths and weaknesses of each of the regulatory approaches as embodied by the GDPR. What must be examined now is how these approaches can best be utilized going forwards. This will involve a comparison of the various advantages and disadvantages of the regulatory approaches and a consideration of how the law, when taken as a whole, can be used to best regulate bots and AI.

There are a number of observations that can be made from Sect. 3. First, where the processing is substantially the same whether it is performed by a bot or by conventional, human processing, the use of general provisions should be preferred. By using general rules, it is possible to create a simpler legal regime, which means that data subjects will be more aware of their rights, and increase the ease of implementation and legal certainty for controllers. In such a situation, the use of a specific rule would provide no additional protection for data subjects; there is, therefore, no reason to

[48] Koops (2014), p. 259.

[49] For a list of companies who have self-certified under the Privacy Shield, see https://www.privacyshield.gov/list Accessed 12 September 2017. For a list of companies who were self-certified under Safe Harbour, see https://www.export.gov/safeharbor_eu. Accessed 12 September 2017.

[50] See, e.g., https://www.iso.org/standards.html. Accessed 12 September 2017.

introduce them. Equally, there is no compelling reason to introduce co-regulatory provisions simply because bots and AI happen to be involved.

One potential criticism of this observation is the general unease surrounding processing by bots and AI. It is not uncommon for certain bots to be described as "creepy."[51] Further, as bots and AI represent the cutting edge of technology and are often extremely complicated, there is a risk that data subjects will be unable to tell the difference between a bot which is creepy or harmful and a bot which is harmless. It could, therefore, be argued that data subject confidence (one of the goals of the GDPR, as stated in Recital 7) would be increased by a shift from general to specific legislation, even if the contents of those rules are substantially the same. This should not be accepted. While it may increase confidence amongst some data subjects, this would not outweigh the disadvantages of creating a more complicated legal system, nor of the risks that the specific legislation diverges from the general rules. Rather, the better solution is for a promotion of education and awareness about bots, enabling data subjects to make more accurate judgments and decisions.

General provisions are also useful where the rules are equally as desirable to both general processing and to processing specifically performed by bots and AI (e.g., the documentation rules in Article 30). In such a scenario, there is no need to complicate the legal regime by creating duplicate rules. Here, it is irrelevant whether the processing causes dramatically different results or uses dramatically different procedures; what matters is that the imposition of the rules or standards is equally desirable whether or not bots and AI are involved. One difficulty with this is that it is not always obvious when the general provisions will be universally desirable. Even if it were possible to state with absolute certainty that a rule or principle is desirable in every situation that currently exists, there is no way of knowing that a novel situation will not arise which fundamentally challenges the underlying justification. While this is more likely in some situations than others, it must be remembered that bots and AI are still a developing technology and it is not always easy to predict how it will develop. Given that the Data Protection Directive[52] was in place for 20 years before the GDPR was enacted, a period which saw a dramatic shift in technological sophistication, we must accept that it is likely that similarly significant shifts will occur during the life of the GDPR.

This, then, is an opportunity for general rules to be complimented by co-regulatory provisions. As these provisions can react more flexibly than the specific rules, it is possible to allow for an "updating" of the law to match recent developments by using co-regulatory tools to provide guidance on how to implement one's general legal obligations. This approach will be more useful when dealing with principle-based general rules (such as Article 5) than rules-based general rules (such as Articles 13–14). However, there are risks with over relying on this approach. Introducing

---

[51] See, e.g., Leonard (2014), p. 57. There is also great public fascination with the idea of "creepy" AI, even where the bots only seem creepy because of misinformation or inaccurate reporting; see Baraniuk (2017).

[52] Directive 95/46/EC of the European Parliament and of the Council of 24 October 1995 on the protection of individuals with regard to the processing of personal data and on the free movement of such data [1995] OJ L281/31.

co-regulation will reduce the simplicity of the general rules, which is one of the large advantages of this approach. Further, this would mean that many balancing and policy decisions are moved from the legislature to the private sector. There is a risk that such private bodies will put commercial interests above those of data subjects and have little incentive (or may even have a disincentive) to go beyond the bare minimum protections as required by law. Where ambiguities exist, there is a risk that co-regulatory tools will exploit them, rather than resolve them in a balanced way. As noted above, this risk is addressed to some extent by the GDPR's approval mechanisms. Even with these mechanisms, however, a legislature coming up with a regulatory scheme should (in theory) place more focus on data subject rights (or should at least make the decision from a more neutral perspective). This criticism applies even more strongly to co-regulatory provisions like DPIAs and Data Protection by Design, where no such approval mechanism exists.

This discussion highlights the advantages of specific rules. Here, the legislature is able to directly address the balancing acts of the different interests involved in certain issues. As the discussion in Sect. 3 demonstrated, a specific regulatory approach would have allowed for a more in-depth discussion on the definition as relates to processing by bots and AI. This, in turn, would allow the law to lay out a more precise rule which carefully balances the interests. It should not, however, be automatically assumed that the use of specific rules will avoid this problem. As shown in relation to Article 22, the use of special legislation may still leave a number of ambiguities in the law. Given the nature of specific regulatory tools, it can be difficult to address these ambiguities with co-regulatory solutions; rather, controllers must often wait for a court to interpret the law before they can have the desired legal certainty. One argument is that this is not an issue with specific regulation per se, but rather an issue with the drafting of a particular specific regulation. While Article 22 is ambiguous, it does, at least, enable those ambiguities to be resolved by a court in a way which focuses directly on bots and AI, rather than requiring the issues to be balanced to a general processing environment. Nevertheless, we must avoid arguments which suggest that while general regulation will usually have to create compromises to cover multiple situations and co-regulation risks the balance being set incorrectly, specific regulation can precisely balance the law—it is merely that specific regulation has a greater opportunity to do so.

Specific regulations are also weaker than general or co-regulatory approaches in relation to changing factual circumstances. References to profiling, for example, were implemented to address particular concerns that were relevant at the time of drafting. However, there is no guarantee that this issue will be considered so important in 15 years time, rendering the references redundant. More dangerous is the risk that the specific regulatory provisions in the GDPR which currently seem beneficial will actually be harmful after technology has moved on. It is difficult to predict whether or not a provision against automated individual decision-making will be helpful in the next decade. Nevertheless, given the slow legislative process in the EU, it will be difficult to change or remove it when the technology advances. This concern can also apply to general regulations (see, for example, the concerns that the general rules in the Data Protection Directive were simply too old to balance the new norms brought about by modern technologies).[53] However, the extent of this will depend on the regulation in question—principle-based general regulations, such as the principle of lawfulness, fairness and transparency are much less likely to become outdated than rules-based general regulations.

A related issue with specific regulation is that, in order to create a specific regime, we must provide firm legal barriers. Without these, specific legal provision would create significant difficulties with legal certainty (see, for example, the difficulties relating to the term "solely" in Article 22). The problem with designing such a provision is that it is hard to create regulations which point to bots and AI in general, rather than having to rely on regulations which are aimed at related, largely overlapping concepts such as "automated processing." As described above, it is very difficult to come up with a universally accepted definition for bots and AI. How, then, do we decide which processing activities will fall under this specified provision and which ones will not? If we attempt to draw a hard and fast legal line, we will inevitably risk creating undesirable errors; too wide and we will end up regulating processing by programs that people would not naturally consider bots, too narrow and we will end up excluding processing by programs that people would consider bots. This would unhelpfully complicate the legal regime, undermine the protection for data subjects and risk impeding the socially-desirable development of bots and AI.

This is not such an important issue with general or co-regulation. While the issue of defining "bots and AI" does not apply at all to general regulation, co-regulation is able to provide much more reactive tools, removing the need for firm legal definitions. In addition to the ability to create regulation which scales a controller's obligations to the facts without necessarily setting that scale in law (e.g., with Data Protection by Design), it is also important to consider the role of expert bodies. Whereas it would be very difficult to introduce a definition of bots and AI into a piece of static legislation, an expert body (such as the Article 29 Working Party/European Data Protection Board) is more likely to find the right place for that line—and, importantly, is able to adapt and change the definition as the technology moves without needing to pass an amending instrument. We have already seen, for example, the Article 29 Working Party exercise this flexibility by altering its interpretation of the word "solely", as discussed above. These general bodies can be responsible for helping to guide the interpretation of both general and co-regulatory rules.

This approach is, of course, subject to criticism. As strongly argued by de Hert, we should be wary of unconditionally delegating power to secondary bodies.[54] Soft-law bodies often lack democratic accountability and may be left to decide extremely controversial topics with no clear correct answer. In such situations, it is arguable that they provide an inappropriate forum for the decision. It is, therefore, questionable how far they should be left to decide matters of policy. This must be balanced against the advantages of having the decision made by an expert body.

More fundamentally, both general and specific regulations have a clear advantage over co-regulatory techniques, namely that they are already enshrined in law and

[53] The GDPR, Recital 6.

[54] See, e.g., de Hert (2016), p. 464.

are easily and directly enforceable. If law and policy makers wish to allow for co-regulatory techniques to become as valuable as their purely legislative alternatives, they must focus on promoting their adoption. If used properly, the techniques included in the GDPR can play an important role in ensuring a balanced and intelligent level of protection (not least because of their ability to scale to the nature of the bots or AI in question). Nevertheless, we must not encourage unrestrained usage. As noted above, co-regulatory tools will inevitably add complication to the legal regime, which should be avoided where possible.

By comparing the relative strengths and weaknesses of the regulatory techniques, it is clear that a balanced approach is necessary. One of the first observations of this chapter was that we cannot assume that any two sets of processing activities involving bots and AI will be the same; it would, therefore, be foolish to assume that any two sets of processing activities could be best regulated by the same legislative techniques. It is suggested that to create a specific law covering processing by bots and AI would be a mistake; while the use of specific regulation will allow for accuracy and a nuanced argument, this would create too convoluted a legal regime given the possible variation in AI and bot activities. However, the general regime must be complimented by other regulatory tools where their strengths and weaknesses make them appropriate.

This is, largely, the approach taken by the GDPR. By mixing the general, specific and co-regulatory approaches, the law allows for a combination of provisions. The question that remains, however, is whether the GDPR gets this balance right—-something which will have to be seen as the law begins to take effect. It should be remembered that, where time shows that general provisions are completely inappropriate for bots and AI, or when it becomes clear that more specific provisions are necessary, it is always possible to pass a dedicated law to cover these issues (as was done with, for example, the e-Privacy Directive).[55] Equally, there is plenty of space in the GDPR for co-regulatory approaches to take a wide view of their mandate if the law does not naturally support a desirable regulatory environment.

As is always the case in law and technology, the regulation can only do so much to keep up with bots and AI, but it is suggested that the GDPR does contain the tools to at least try. The general regulations will provide enough space for the law to deal with changing realities; the specific regulations which exist tackle the problems that were seen at the time of drafting; and the co-regulatory approaches provide the opportunity to provide a flexible, but structured, regime. What is important is that the law is interpreted in a sensible way which applies general provisions in a way which makes sense, which interprets specific provisions carefully so as to apply the intended balancing of interests and which encourages private parties to use co-regulatory approaches in a way that allows them to make judgment calls in an informed manner.

---

[55] Directive 2002/58/EC of the European Parliament and of the Council of 12 July 2002 concerning the processing of personal data and the protection of privacy in the electronic communications sector (Directive on privacy and electronic communications) [2002] OJ L201/37.

5 Conclusion

This chapter began by suggesting that we must avoid overstating the difficulty of the task with which the law must overcome. There is no need to regulate a perfect, human-like AI. The challenge that this chapter then posed, however, may actually be just as difficult. Instead, we must regulate a cutting-edge technology which has the potential to perform acts which can be either revolutionary or trivial. To make matters worse, this technology is almost impossible to define concretely and is a mystery to a great number of people, but is already found in countless products and services. Meanwhile, the law which attempts to govern it must deal with all kinds of processing of personal data, from the simple act of taking a register of attendance to complicated systems designed to predict a person's every move and preference.

How, then, does the law attempt to deal with this problem? This chapter identified three broad regulatory techniques: the general, the specific and the co-regulatory approaches. The fundamental conclusion of this chapter is that each approach has certain strengths and weaknesses. When used correctly, these approaches can be used in a way that compliments the others, with general regulation providing an easily understandable regime, specific regulation dealing with the topics which cannot fit into the general regime and co-regulation providing a supporting role. The GDPR already attempts to implement these approaches in this way, although it will have to be seen how successfully it has done so in practice.

Having answered this question, we can conclude that the law has, at the very least, taken a correct approach to drafting. What must be seen now is whether the actual content of the GDPR will adequately regulate the issues surrounding bots and AI. At this stage, it is difficult to provide any reliable prediction; the law has not yet come into force and bots and AI are still developing at a rapid pace (in often unexpected directions). It is hoped that the observations made above will, nevertheless, provide encouragement. There is no need to fundamentally reinvent the approach to the regulation of data protection, merely a need to ensure that that approach is being used in an intelligent and sensible way.

Acknowledgements This chapter was written as part of the Future Regulation of Industrial Internet project at the IPR University Center, Helsinki. I wish to express my gratitude to Olli Pitkänen for comments on an earlier draft of this chapter.

References

Article 29 Working Party. (2011). *Opinion 15/2011 on the Definition of Consent.*

Article 29 Working Party. (2017a). *Guidelines on automated individual decision-making and profiling for the purposes of regulation 2016/679*

Article 29 Working Party. (2017b). *Guidelines on Data protection impact assessment (DPIA) and determining whether processing is "likely to result in a high risk" for the purposes of Regulation 2016/679*

Article 29 Working Party. (2018). *Guidelines on Automated individual decision-making and profiling for the purposes of regulation 2016/679* (revised).
Baraniuk, C. (2017). *The 'Creepy Facebook AI' story that captivated the media*. BBC News, August 1, 2017.
Binns, R. (2017). Data protection impact assessments: A meta-regulatory approach. *International Data Privacy Law, 7*(1), 22.
Bygrave, L. (2001). Automated profiling: Minding the machine: Article 15 of the EC data protection directive and automated profiling. *Computer Law & Security Report, 17*(1), 17.
Council of Europe. (2010). *The protection of individuals with regard to automatic processing of personal data in the context of profiling: Recommendation CM/Rec(2010)13 and explanatory memorandum.*
Council of the European Union. (2014). *Right to be forgotten and the Google Judgment*. Interinstitutional File 2012/0011 (COD).
de Hert, P. (2016). The future of privacy. Addressing singularities to identify bright-lines that speak to us. *European Data Protection Law Review, 3*(4), 461.
European Commission. (2012). Proposal for a regulation of the European Parliament and of the Council on the protection of individuals with regard to the processing of personal data and on the free movement of such data (General Data Protection Regulation). COM(2012) 11 final.
European Parliament. (2011). Report on a comprehensive approach on personal data protection in the European Union 2011/2025(INI). A7-0244/2011.
Information & Privacy Commissioner of Ontario. (2013). *Privacy by design*. Available at: https://www.ipc.on.ca/resource/privacy-by-design/. Accessed December 8, 2017.
Kamarinou, D., Millard, C., & Singh, J. (2016). *Machine learning with personal data*. Queen Mary Legal Studies Research Paper 247/2016.
Koops, B. J. (2014). The trouble with European data protection law. *International Data Privacy Law, 4*(4), 250.
Leonard, P. (2014). Customer data analytics: Privacy settings for 'big data' businesses. *International Data Privacy law, 4*(1), 53.
Maxwell, W. (2015). Principles-based regulation of personal data: The case of 'fair processing'. *International Data Privacy Law, 5*(3), 205.
Mayer-Schönberger, V., & Cukier, K. (2013). *Big data: A revolution that will transform how we live, work and think*. John Murray, Kindle edition.
Míšek, J. (2015). Consent to personal data processing—The Panacea or the dead end? *Masaryk University Journal of Law & Tech, 8*, 69.
Nwana, H. (1996). Software agents: An overview. *Knowledge Engineering Review, 11*(3), 1.
Obar, J., Oeldorf-Hirsch, A. (2016). The biggest lie on the internet: Ignoring the privacy policies and terms of service policies of social networking services. In *TRPC 44 Conference on Communication, Information and Internet Policy*, Virginia, 2016. Available at: https://papers.ssrn.com/sol3/papers.cfm?abstract_id=2757465. Accessed January 3, 2018.
Parker, C. (2007). Meta-regulation: Legal accountability for corporate social responsibility. In D. McBarnet, A. Voiculescu, & T. Campbell (Eds.), *The new corporate accountability: Corporate social responsibility and the law*. Cambridge: Cambridge University Press.
Pearce, H. (2015). Online data transactions, consent and big data: Technological solutions to technological problems. *Computer and Telecommunications Law Review, 21*(6), 149.
Rubinstein, I. (2013). Big data: The end of privacy or a new beginning. *International Data Privacy Law, 3*(2), 74.
Russell, S., & Norvig, P. (2016). *Artificial intelligence: A modern approach*. Upper Saddle River, New Jersey: Pearson.
Sartor, G. (1993). *Artificial intelligence and law: Legal philosophy and legal theory*. Norweigian Research Centre for Computers and Law, CompLex 1/93, Tano, Oslo.

I, Inhuman Lawyer: Developing Artificial Intelligence in the Legal Profession

Dena Dervanović

Abstract What are the possibilities of having AI lawyers in the true sense—as autonomous, decision-making agents that can legally advise us or represent us? This chapter delves into the problems and possibilities of creating such systems. This idea is inevitably faced with a multitude of challenges, among them the challenge of translating law into an algorithm being the most fundamental for the beginning of the creation of an AI lawyer. Moreover, the chapter examines the linguistic aspects of such a translation and later moves on into the ethical aspect of creating such lawyers and ethically codifying their conduct. This is followed by a brief deliberation on whether Asimov's Three Laws of Robotics would be helpful in this regard. The ethical discussion results in a proposal for a concept of Fairness by Design, conceived as the minimum standard for ethical behavior instilled in all AI agents. The chapter also attempts to give a general overview of the current state-of-the-art AI technologies employed in the legal domain as well as imagines the future of AI in Law. Subsequently, the chapter imagines an AI agent deal with and resolve the "Solomon test" of splitting the baby. Finally, it is concluded that the advantage of having AI lawyers can be measured by the possibility of redefining the legal profession in its entirety as well as making legal advice and justice more accessible to all.

Keywords AI · Robots · Lawyers · Legal profession · Ethics

1 Introduction

> But the man is quite inhuman, Dr Lanning.[1]

The man in question is Stephen Byerley, a district attorney suspected of being a robot—never seen sleeping or eating, always perfectly composed and incredibly good

[1] Asimov (1950), p. 189.

D. Dervanović (✉)
Stockholm, Sweden
e-mail: dervanovic.dena@gmail.com

M. Corrales et al. (eds.), *Robotics, AI and the Future of Law*, Perspectives in Law, Business and Innovation, https://doi.org/10.1007/978-981-13-2874-9_9

at his job. One could argue these are all features that instill a degree of fear and envy in an average human being. The story of Stephen Byerley will be referenced throughout this chapter, since Isaac Asimov has long been an inspiration for me. Asimov has famously fiddled with the concept of robotics, AI and the role they will play in the future of humanity. Will they be confined to the limits of being our smart home appliances and modern-day slaves or will they amount to more than that—will they be our doctors, soldiers, CEOs or lawyers? Will we allow them to? More importantly, should we? Note the tone of the word "they" used in these questions. It almost rings with anxiety and fear, alluding to "them" being an imminent danger of our own creation, yet another man-made disaster waiting to happen. Before we put our tinfoil hats on in the search of adequately catastrophic reasons not to indulge ourselves into "relinquishing control" in favour of artificial intelligence, we should try to examine the probability of this ever happening and assess the possible consequences and ways to get to such a technologically advanced stadium.

The concern of this chapter will remain in the legal and ethical domain, with its focus split in two—one part is focused on legal semiotics and the other on ethics. With legal semiotics, we embark on an exploration of whether law is a discipline that we can translate to an algorithm. Can we transpose the contextual or linguistic nuances of the law into an algorithm? Should we trust an AI lawyer to review our contracts, or give legal advice or representation? It does not require long and hard thinking to assume that the technology required for doing at least some of these things is not too far off in the future. In fact, some law firms are already doing document review with the help of AI, improving efficiency and letting lawyers do more intellectually stimulating legal work.[2] On the other hand, the vision of a fully autonomous AI lawyer is a bit further down the road and this idea will initially require some analysis to determine whether law, a millennia-old discipline and a true ars antiqua, is construed in a way to support the notion of an inhuman lawyer?

Furthermore, we will tackle the question of ethics: what is at stake here? Do we need to employ ethical and moral rules into an AI lawyer? With the increasing use of autonomous decision-making, how do we assure that an AI lawyer would be capable of making moral and ethical decisions? Law is, after all, a discipline based on values such as fairness and equality.[3] How does this fare against AI bots and autonomous decision-making agents we've seen thus far? Think about Google not showing executive jobs to women, or its Google Image Search identifying black people as gorillas,[4] or the Microsoft Twitter bot that had to be taken down after merely 24 h online because it was displaying racist, sexist and anti-Semitic behaviour and overall bias.[5] The whole point of AI is to create systems that have a degree of autonomy and an ability to mimic intelligent human behaviour, and this emphasises the necessity to introduce ethical rules for that behaviour not to adversely affect people or worse, completely slip out of control.[6] We have already seen threats to the

[2] Weller (2016).

[3] Kennedy (2017), p. 170.

[4] Kasperkevic (2015), quot. in. Kennedy (2017), p. 172.

[5] Kennedy (2017), p. 172.

[6] Anderson (2011), p. 294.

rule of law and justice when using autonomous decision-making systems, e.g., in California where people got wrongly arrested and registered as sex offenders.[7] The way justice works is that it is impartial, objective and its ultimate goal is to be fair. Therefore, it is important to take a look at the manner in which we are developing the AI and Law domain—and it is vital that we ensure that autonomous decision-making systems are not too flawed in the beginning, to ensure that we can still do something about those flaws. In this sense, I would like to draw an analogy with the principles of Privacy by Design[8] in data protection law—i.e., privacy principles and settings should be incorporated at the very beginning stage of the development of a tool, service, or product in order to ensure a holistic manner in which to safeguard the user's privacy and adhere to data processing principles.

Along the same lines, we ought to ensure to employ "Fairness by Design" and put special effort in creating algorithmic solutions that will not discriminate on any basis, and this should be done now, while we are still at a fairly rudimentary stage of the AI and Law domain. Of course, not all is already lost—there are some present examples of advancements in creating autonomous decision-making systems where these systems have not been detrimental to people they targeted: systems such as DoNotPay.co.uk who have helped people handle their parking tickets as well as refugee status applications.[9]

A note to the reader: this chapter poses a lot of questions and it would be wise not to expect answers to all of them. The aim of this chapter is to explore possibilities, compare theories and embark on a visionary journey for the future of law and the legal profession. One of the reasons why it is important to embark on such explorations is because major AI advances have, thus far, avoided the legal domain. There may be several factors that contributed to this: the fact that AI researchers are usually not lawyers or legal professionals and therefore lack the interest and knowledge to employ law in their development of AI systems and technologies, and the fact that lawyers have not exactly been receptive to the notion of using emerging technologies, and specifically AI, in their work.[10]

Furthermore, this chapter envisions an autonomous decision-making machine, i.e., agent, such as Mr. Byerley, practicing law to the fullest extent of it and this should be borne in mind when reading. An autonomous machine is "a system situated within and a part of an environment that senses that environment and acts on it, over time, in pursuit of its own agenda and so as to effect what it senses for the future."[11] The elements of this definition are the following: this machine is "reactive, self-controlling, goal-oriented and temporally continuous."[12]

Therefore, this chapter is trying to look into a fairly distant future where we may be surrounded by many Mr. Byerleys and where the legal profession will look

---

[7]Farivar (2016) quot. in. Kennedy (2017), p. 172.

[8]Cavoukian (2010).

[9]Fingas (2017) quot. in. Kennedy (2017), p. 171.

[10]See also Mommers et al. (2009).

[11]Frankling and Graesser (1997), quot. in. Brożek and Jakubiec (2017), p. 293.

[12]Frankling and Graesser (1997), quot. in. Brożek and Jakubiec (2017), p. 294.

somewhat different to what it looks like today. Note that the terms "machine," "AI," "robot," "agent" and "system" are used interchangeably. Moreover, this chapter does not deal with the legal responsibility of AI lawyers. This chapter, instead, may be viewed as preparatory work for the question of legal responsibility of AI lawyers and AI agents, in general.

This chapter does not pretend to be much more than what it was intended to be: a thought experiment about the future of the legal profession without the apocalyptic connotations that often characterize this line of thinking. It is but a way to imagine the future and our role in it; how we can help shape it so that it turns out for the better rather than for the worse.

Lastly, if we are discussing AI lawyers, why not mention the possibility of AI judges? In the last section, a discussion on how an AI judge would handle the Solomon test is presented. I argue that, while having AI lawyers may be a thing of the future, having AI judges will require a longer waiting time and much more sophisticated technology.

2 Codification-to-Code

This chapter delineates the importance of diving into what the law is, all with the aim of identifying the way in which law can be found translatable into algorithms. This is the founding premise of creating a fully autonomous AI lawyer that is able to practice law. This chapter discusses the importance of linguistics and the challenges of translation. The nature of law must be examined in order to conduct a profound analysis of the question of inhuman lawyers. The main conundrum here is whether law is a mathematical formula ready to be transposed into an algorithm or is it perhaps something more than that—a system incapable of being reduced to an algorithm purely because of the importance of its interpretation methods and circumstances to be applied to each specific case?

The answer to this future-oriented question may lie in existing legal theory and semiotics—yet again proving that law, as a discipline, is fixated on digging through its past, be it with finding case law precedents or travaux preparatoires of a legal instrument. As a domain that is necessary to establish order in society, law is an essential pillar in any society governing nearly every aspect of human life, thereby creating generally adopted definitions of everyday concepts.[13] Considering its extensive history and its many nuances, the legal system in any country will undoubtedly be a large body of rules and regulations intrinsically intertwined.[14] This makes it all the more difficult for anyone without a legal education to access relevant and comprehensive legal information about a subject matter.[15] In addition, it is nearly impossible to expect that a single lawyer would be able to know all the existing laws

[13] Lame (2004), p. 382.

[14] Bibel (2004), p. 163.

[15] See Mommers et al. (2009) p. 52.

within a system and to be able to know their content and application. As such, it does not take long to assume that the legal system (in general) will, in and of itself, present various challenges in the way of our idea of translating law into a set of algorithms.

There is an argument by McGinnis and Wasick that I would like to take note of here. This argument describes law as an information technology of sorts—it distributes information about legal norms to the public and is fed with information from the public about what those norms should contain and ultimately creating change to those norms.[16] This means that this approach sees law as an information technology working two ways: top-down, representing the norms being communicated to society, and bottom-up, representing society's feedback and reaction to such norms.[17] The normative framework is indeed intended to inform society about what is permissible and what isn't, thereby regulating society through information provision.[18] This is a fairly good representation and summary of the way law is intended to function and has functioned since ancient Roman codexes. This ultimately means that, in today's society, the channels of communication of legal norms are essentially an interdisciplinary exercise: one that combines law as the subject matter and modern information technologies as a means for conveyance of the legal norms to the public.[19]

Richard Susskind echoes this in a way, by saying that we have evolved in the ways we convey information, legal or otherwise, and that we are currently experiencing a fundamental shift in how information flows in our society.[20] This shows no signs of stopping, and transcends disciplines and industries with an ease that, if it weren't impressive, would most likely be frightening. Why would law be any different—lawyers, judges and legal systems all around the world should be, and some have started, making use of developing technologies to ensure more efficient and thorough work and are, in the case of law firms, hopefully moving away from the traditional model of billing by the hour.[21]

2.1 Linguistic Considerations

Let us explore the way this vision of an AI lawyer could come about. Most new technological breakthroughs in the realm of AI rely on methods such as natural language processing, like for example Siri, Alexa or Cortana. This chapter argues that natural language processing will prove to be essential to the advancement of the AI and Law domain, both in the way of creating AI lawyers but also in creating systems that will help the general population understand the law to a greater extent.

---

[16] McGinnis and Wasick (2015), p. 993.

[17] McGinnis and Wasick (2015), p. 997.

[18] Bibel (2004), p. 164.

[19] Mommers et al. (2009), p. 53.

[20] Susskind (2017), p. 191.

[21] See also Susskind (2017).

Natural language processing, contrary to key word searches provide with a more "spot on" result of the search, bringing the searcher closer to their answer by trying to understand the meaning of the search and thus making the search experience more intuitive and efficient.[22] Over the course of the past decade, we have seen a significant improvement in the domain of natural language processing and its ubiquitous use.

Natural language processing is defined as "a computer program's ability to understand spoken and written language and is a component of Artificial Intelligence."[23] Because of what it entails and what it can do for AI in general, it is not a stretch of imagination to say that natural language processing is going to prove vital to the development of AI lawyers. If we take the example of IBM's Watson, we can conclude that the strength of natural language processing is impressive to say the least. Watson functions on the principle of finding the most relevant and accurate sets of answers to a question, and then making a risk assessment in order to provide the final answer.[24] IBM has asserted that this method could be employed for legal documents as well[25] and this will bring about an inarguably positive change in the way legal search works, by making it more efficient and intuitive. This makes natural language processing a valuable tool for creating and developing AI lawyers, simply by virtue of what this kind of processing entails—a smoother, efficient and more focused way of searching for legal information, be it in the form of a norm or case law. No matter the level of its formality, legal language is, in fact, natural language, i.e., it lacks its own linguistic rules.[26] Therefore, translating this language could very well be in the scope of natural language processing. We have seen valuable progress when it comes to "normal" language processing, but the true measure of how far we can go is when this processing turns to technical terminology and language of domains such as law. In this sense, we will be dealing with a perfect understanding of precise terms, their context and exceptions if any.[27]

An experiment aimed at testing how natural language processing interacts with the law was done on the French body of law (Codes). This experiment involved identifying legal concepts by performing syntactical analysis on the text—thereby identifying nouns, verbs, adjectives and later on more complex constructs.[28] By employing methods of identification of terms, syntactical analysis, making semantic relationships and with the help of some human intervention, researchers were able to create a helpful set of techniques using natural language processing to help ontologists with identifying concepts—thus making natural language processing a tool in

---

[22]This principle is applicable in legal search as in any other. See also McGinnis and Wasick (2015), p. 1017.

[23]Bouaziz et al. (2018), p. 2.

[24]IBM Watson. Available at: https://www.ibm.com/watson/. Accessed 27 April 2018.

[25]IBM Systems & Technology Group (2011), p. 5.

[26]Mommers et al. (2009), p. 55.

[27]See Wettig and Zehendner (2004).

[28]See Footnote 13.

creating ontological systems that can be used in information retrieval processes in AI systems of the legal domain.[29]

This is a good start and probably a fundamental one in creating an AI lawyer. However, here is where we face a number of challenges, with the most obvious one glaring into our faces: legal search is merely one of the first phases of answering a legal question. In this sense, we can imagine a powerful supercomputer being able to provide us with the most relevant set of answers to a legal conundrum, but we can safely assume that it will still need to be handled by legally trained staff in the near future, in order for them to ask the right questions and lead the computer in the right direction, and that the final answer to the posed conundrum would have to be deduced by the (human) lawyer herself, based on the spectrum of answers provided by the supercomputer.

2.2 *The Risk of Getting "Lost in Translation"*

Even if fundamental, natural language processing will arguably face setbacks in its attempt to create and develop reliable, capable AI lawyers. At least at first. In other words, it just might not be enough. This is rooted in the theory of "the law of interpretation"[30] which relies on determining what legal instruments mean in a given legal system without necessarily relying on linguistic interpretation, but rather a legal one, i.e., the relationship between different norms, contexts and overall a balance between all the elements of the subject of interpretation must be taken into account.[31] This is strongly linked to the fact that legal interpretation will often seek to read between the lines i.e., find implicit meaning in a legal text, whereas linguistic interpretation takes the text at face value and in its entirety.[32] I will try to summarize this theory in a few sentences and apply it to the essence of this chapter. Fundamentally, what this means is that understanding the law of interpretation is to be able to discern language from law.[33]

In other words, language is a tool used in the expression of law, but it can only go so far when interpreting the law. If this weren't the case, we would not need trained professionals to interpret or write legal codifications and texts because a linguist might as well be able to do it, which is not exactly the reality of legal writing and interpretation. Here we can distinguish how the two disciplines treat validity: generally, an unused legal provision may still be valid in the formal sense, i.e., time or lack of practice does not necessarily diminish its validity, whereas language generally is shaped by its usage and practice, meaning that expressions, rules and other elements

[29] Lame (2004), p. 395.

[30] Baude and Sachs (2017), p. 1085.

[31] Baude and Sachs (2017), p. 1083.

[32] Husa (2016), p. 263.

[33] See Footnote 30.

may disappear over time if unused.[34] Furthermore, because of a distinction in how language is used in everyday life versus the way language is used in law, we can conclude that legal language (and terminology) greatly differs from its everyday counterpart. This form of "semi-artificial natural language"[35] is characterized by a greater presence of precision and conditioning compared to everyday language.[36] This, however, does not preclude legal language falling victim to the same perils that regular language faces, such as misinterpretation and subjectivity as well as poorly constructed sentences.[37] This goes to show that complex grammatical constructs are but one of the obstacles that we would be faced with at first in our attempts to create a fully functioning robot lawyer.[38] Furthermore, legal language, no matter how precise and distant from natural language, is, still, natural language in and of itself, therefore it greatly differs from the programming language we aim to translate it to.[39]

There is an abundance of material dissecting the relationship between translation and interpretation when it comes to legal texts, and this alone suggests that there is no consensus on how to do this in the best way. It is no secret that the relationship between translators and lawyers is not exactly a walk in the park. This brings us to the following question: why bother with translation and interpretation at all? How does this relate to a robot lawyer and our possibility of creating one? The short answer to this is that, ideally, a fully-functioning robot lawyer will be able to deal with comparative law, legal translations and the complexity of discerning between legal systems with an ease that human lawyers might not possess.

Different legal systems carry different interpretive rules[40] and it is worthwhile to note that interpretation of legal texts differs greatly depending on the region, or legal system. I will use an example provided by Husa: the European Court of Human Rights uses the so called "dynamic interpretation" in order to interpret the European Convention on Human Rights; meaning that they do not necessarily delve into the intent of the drafters in order to understand the meaning of the Convention's provisions.[41] Therefore, the question remains open: how would an AI lawyer be able to tackle this? Or should we take a conservative position and just assume that an AI lawyer would always be limited to one (or a few similar) legal system?

In this sense, and in an attempt of an unbiased general assessment, one could say for example that the continental law system and the common law system may greatly differ in the way they will be translated into an AI system capable of processing and understanding them. The reason behind this is that the continental law system has more structural boundaries, i.e., its format and procedure for creating the law, as well as interpreting and enforcing it is more limited than its common law

[34] Baude and Sachs (2017), p. 1123.

[35] See Footnote 14.

[36] See Footnote 14.

[37] See Footnote 14.

[38] See Footnote 10.

[39] See Footnote 26.

[40] Baude and Sachs (2017), p. 1088.

[41] Husa (2016), p. 270.

counterpart. This difference greatly relies on the nature of precedent and the role of case law. One could argue that the continental system would have the upper hand here due to the fact that case law plays a significantly smaller role in the creation of law. However, if we flip the coin, precedent is generally used as a specific solution to a specific legal problem, which may in fact result in an enabling circumstance for translation into algorithms because it is a more specific, mathematical solution to the given problem. However, either system is becoming slightly more like the other, and this is unlikely to stop happening in the future. Regardless, all of this is very debatable, and it will become relevant for the creation of AI lawyers eventually.

However, we needn't go too deep into the trenches of different legal systems or interpretation rules at this point in time in order to identify that an AI lawyer that is able to execute proper interpretation of legal norms and texts can be a challenge to create and that any poor or mediocre execution would carry the risk of unfavorably affecting those who relied on it. The crux of most of techniques and methods for translating law is based on human intervention, both in the form of support and upkeep and will likely remain so for the foreseeable future.[42]

2.3 Is Law Ready to Become an Algorithm?

In answering the question of whether law can be made into an algorithm, we delve into the question of whether mathematics and law can be treated as two sides of the same coin. Let us spend a moment untangling this. Please note that the following is a simplified view of mathematics as a discipline. Mathematics is consisted of rules and exceptions to those rules. It is riddled with theories, theorems and problems to solve. Aristotle was of the view that mathematics is defined by the way it studies things (rather than the content of the things it studies) and its "*degree of abstractness*."[43] Mathematics is essentially a study of hypothesis and its substance with drawing the "*necessary conclusions*."[44] Mathematics, as Kac and Ulam pointed out, "settles a question, suggests new ones, and leads to new observations."[45]

For a successful solution to a mathematical problem, one needs to employ the said theories by knowing how to discern them, one needs to apply the rules and employ mathematical logic.[46] The answer to any mathematical problem has one out of two outcomes: the correct one and the incorrect one. That is one of the perks of being an exact science, i.e., the reliability of its outcomes. In very simplistic terms: if the numbers are off, the problem remains unsolved.

---

[42] See Footnote 19.

[43] Moore (2010), p. 4.

[44] Moore (2010), p. 7.

[45] Kac and Ulam (1968), p. 4.

[46] Hodel (1995), p. 1.

On the other hand, lawyers are known for their general dislike of anything to do with mathematics.[47] Jokes aside, let us look at law: for a successful solution to a legal problem, one needs to apply existing rules that govern it, one needs to employ relevant theories and know how to discern them, rules and procedures of interpretation, examine the stare decisis and lastly, one must employ logic in order to reach a conclusion through syllogism i.e., an inference of a major premise, a minor premise and a conclusion. Therefore, to assume that law is merely a science of rules is to be gravely mistaken about the nature of law. In fact, law is usually inaccurately perceived as a rule-based discipline, mostly by the general public due to a misconception that has been in the mainstream for a while now—completely neglecting the fact that argument plays a big role in law.[48] It has initially been thought that legal reasoning is an axiomatic system, much like mathematics, meaning that rules equal axioms and their analysis was therefore based on deduction.[49] This was later proven to be incomplete and moderately inaccurate—legal reasoning encompasses a lot more and the logic deployed in legal reasoning is not as black-and-white in the sense that it only contains deduction.[50] This, arguably, makes things a tad more difficult for the translation of law into algorithm.

Compared to mathematics, there may be many a solution to a legal problem, because of all of the variables that may come into the problem's features and circumstances. Hence, there may be a set of solutions to a legal problem, and the discipline allows for different perspectives and this discussion can be solved in more than one way.[51] The fact that law is, to put it mildly, indeed a challenge to translate into algorithm has been confirmed by other researchers on the topic.[52] These difficulties ultimately lie in the power that argumentation holds in law: disputes are, after all, resolved by presenting a more compelling argument for the case and in line with the legal rules governing the matter.[53] Furthermore, law belongs to social sciences whereas mathematics belong to natural sciences and there is where we see perhaps the largest difference (and difficulty): while natural sciences and disciplines may be quantified, they also possess a certain level of "detachment"[54] that cannot be found in social sciences because they are much more related to a certain system where they originate from and are dependent on social circumstances in a particular setting thereby making them more difficult to standardize and "detach" from the reality where they originated from and operate in.[55] This ties in with the differences between legal systems mentioned above, as well as with differences in interpreta-

[47] Henket (2003), p. 1.

[48] Leith (1988), p. 32.

[49] Bibel (2004), p. 176.

[50] See Footnote 49.

[51] For the sake of the research questions, it is vital to know that this paper uses a generalized view of the discipline of law.

[52] Leith (1991), p. 201.

[53] Trevor et al. (2006), p. 1.

[54] Leith (1988), p. 34.

[55] See Footnote 54.

tion and, when it comes to law, the intentions of the legislator and the spirit of the legislation as well as the social context in which it is written.[56]

The closest we have come to such translations of legal norms to formulas is in the field of formal deontics where we study legal norms by using mathematics.[57] This creates an opportunity to achieve the matter of translation we discuss here but it is immensely challenging. Seeing as the nature of a legal norm is one of an "open textured concept"[58] which signifies that they are not to be taken at face value, i.e., different factors will impact the interpretation and definition of the norm.[59] We can use a plethora of regular, basic legal concepts to prove that context is key in this regard—even classifications of legal instruments can vary from jurisdiction to jurisdiction, whereas some legal concepts carry a different meaning depending on the contextual relationship they hold with other elements.[60] However, it has been argued that mapping legal concepts, as such, could be achieved by using natural language processing techniques—the technology is mature enough for that; this would be achieved with the help of legal glossaries and lexicons.[61] This is arguably a large body of work, and it would most likely have to be done jurisdiction by jurisdiction.

3 Machine Ethics

If we optimistically assume that the challenges discussed above are resolved and that both law and technology allow for an AI lawyer to exist in the most technically possible and unobstructed manner, we can move on to the last remaining challenge in creating an AI lawyer: the one of ethics. This chapter explores the emerging field of machine ethics and in doing so, it gives mention to the different aspects of discourse and debate in this field and shines a light on why this is of importance. A reference to Asimov's laws of robotics is made in an attempt to prove that they do not suffice, although the idea may be a good beginning.

Obviously, the machines discussed in this chapter are nowhere close to being developed and this idea is probably better fit for a Hollywood movie in the short term. This does not, however, mean that this is not where we are headed in the long term: fully autonomous, decision-making machines working as lawyers, doctors, you name it. If a machine is to make decisions and be autonomous, the machine will need standards and rules of ethics and morals to guide it.[62]

Finally, a concept for tackling all AI systems that is based on foundational ethical principles is presented, albeit only as a concept. There is yet another challenge to

[56] Leith (1988), p. 35.
[57] Stolpe (2010), p. 247.
[58] Lame (2004), p. 380.
[59] See Footnote 58.
[60] Mommers et al. (2009), p. 72.
[61] Mommers et al. (2009), p. 75.
[62] Wallach and Allen (2008), p. 23.

face when imagining a machine capable of ethical behavior: the one of engineering. How do we make this happen in the technical sense?[63] A debate on this topic must be started, sooner rather than later.

3.1 Machine Ethics as a Field of Research

The research in the field of machine ethics is mainly focused on "the possibility of constructing machines that can mimic, simulate, generate or instantiate ethical sensitivity, learning, reasoning, argument, or action. The machines in question may be physical or virtual; they may be stationary or mobile."[64] Nowadays when we talk about machine ethics, we mostly discuss scenarios of self-driving cars in trolley car-like scenarios usually taught in ethics classes.[65] Autonomous (viz. self-driving) cars have been at the center of a new wave of ethics discussions—most coming from the angle of trolley-car problems. Faced with a trolley-car scenario, we are to make a choice between how many people get to die as a result of our intervention.

The way this ethical discussion has evolved in the advent of autonomous cars is that the trolley-car problem is not representative of the ethical and moral dilemmas we face when programming autonomous cars.[66] Nyholm and Smids rightly observe that, unlike with trolley-car problems where there is certainty in the outcome, what an autonomous car faces in an event of an imminent accident almost never has a certain outcome.[67] Furthermore, instead of making a split-second decision (as the subject in a trolley-car problem would), the autonomous car would be employing pre-determined and installed ethical codes that would regulate its behavior in a situation of imminent danger.[68] This particular digression from our robot lawyers was done to illustrate the many ways in which we are yet to examine the ethical and moral aspects when it comes to a wide range of AI-related inventions, AI lawyers being but one of the subjects of such deliberations.

The whole purpose of machine ethics is to find a way for AI to be used in a manner that is safe for all humans.[69] This means that machine ethics are there to make sure that no humans are harmed by the use of AI machines—i.e., machine behavior ought not to be detrimental to human kind.[70] The notion of being able to implement ethical rules in a machine to the point where this machine can perfectly execute tasks and make decisions all the while being perfectly in line with ethical codes is so attractive,

---

[63] Wallach and Allen (2008), p. 17.

[64] Guarini (2013), p. 213.

[65] Wallach and Allen (2008), p. 13.

[66] Nyholm and Smids (2016), p. 1288.

[67] Nyholm and Smids (2016), p. 1285.

[68] Nyholm and Smids (2016), p. 1280.

[69] Shulman et al. (2009), p. 1.

[70] Anderson and Anderson (2007), p. 1.

perhaps because humans, as a species, have never been able to do that perfectly.[71] We, as humans, have an abundance of feelings and egotistic behavior that, more often than not, impede us from behaving according to rules of ethics that we may very well be aware of.[72]

Historically, humans have exhibited fear of the unknown or unfamiliar, resulting in irrational and hostile behavior towards those unfamiliar elements; this, for example, led to severe racism.[73] Humans, more often than not, make selective use of ethical rules based on whether it suits them or not in the given moment, thus, being driven by emotions may be detrimental to the application of ethical and moral principles we may well be aware of. Therefore, it may even be appealing that robots would be in charge of certain ethical decisions—devoid of emotions and the human degree of agency, they can apply ethical rules with higher precision.[74]

3.2 Ethical Rules for Robots—Could We Use Asimov's Laws of Robotics?

It has been widely acknowledged that robots are already being developed in what we may call a "faulty" manner—meaning that they do, in their current state, have the ability to create a negative impact on human kind purely due to the fact that machine ethics as a discipline has not fully developed yet.[75] There has already been mention of several examples of this (see above). When thinking about an ethical framework for machines to use, a sci-fi fan's mind will go straight to Asimov's "Three Laws of Robotics:"[76]

"First Law: A robot may not injure a human being or, through inaction, allow a human being to come to harm.
Second Law: A robot must obey orders given it by human beings, except when such orders conflict with the First Law.
Third Law: A robot must protect its own existence as long as such protection does not conflict with the First or Second Law."

Intriguing and widely debated, these laws more often than not create an initial illusion that they could work. In fact, they have been so influential that the invention of the first industrial robot was inspired by Asimov's work.[77] The laws provide readers

[71] See also Anderson and Anderson (2007), p. 4.

[72] See also Anderson and Anderson (2007), p. 2.

[73] Anderson (2011), p. 287.

[74] Anderson and Anderson (2007), p. 4.

[75] Anderson and Anderson (2007), p. 1. The example used here is one of care-robots being developed for elderly-homes in the United States and emphasis is placed on the need to instill ethical principles in those robots to ensure proper care and safety.

[76] Asimov (1950).

[77] Clarke (2011), p. 256.

with a comforting feeling of security—after all, the robotic subjects of the Laws are subordinates to humans and human kind is, seemingly, unendangered.[78] However well these laws fit into Asimov's works, they are unlikely to suffice in the event that we have fully capable AI machines in the way he imagined them in his opus. Asimov himself admitted to a high level of ambiguity in the Three Laws that allowed him to explore those loopholes in his works of fiction.[79] This is widely confirmed by machine ethics researchers and the Laws are deemed inadequate in real-world application, in fact, Asimov uses his fiction to explore the various vulnerabilities of his own laws.[80] However, these Laws still have a strong impact on the field of robotics in general.[81] In fact, there is an immense amount of literature discussing these Laws as a potential departure point, and this chapter is clearly one of them.

Our experience with AI thus far has been rather disappointing with regards to ethics and morals. While we do not have the chance to test AI lawyers for ethics just yet, we have seen bots and autonomous systems that have exhibited intensely discriminatory behavior. When creating rules for the conduct of AI agents, we must ensure that these rules effectively control their behavior and that there is little space for interpretation but, at the same time, we must ensure that the AI agent has the technical capacity to process the laws and apply them in the intended manner—thereby diminishing the risk of misapplication.[82]

Finally, it is here that we realize how important our first question (above) was: can we create laws into algorithms without the risk of having them incorrectly applied or omitted altogether? The question of whether entire legal systems can be transposed into an AI system without consequences for the content and intent of the law extends into whether we can, in fact, implement laws for the AI systems in order to ensure compliance with ethical and other rules that we impose on them. If we know that Asimov's laws are flawed, then we must ensure to find ways of effectively incorporating legal and ethical rules into algorithms, which, as elaborated on above, will prove to be a challenge.

3.3 Why Bother Creating Ethical Constrictions for AI Lawyers at All?

Ethics and morals are an invaluable part of the legal profession and a lawyer is trained to employ them in their everyday work. It has been instilled in the profession; this means that pure legal research in terms of having the ability to have an autonomous system comb through vast amounts of data in a fraction of the time it would take a human in order to obtain specific results is a welcome perk of the job but it will only

[78] Clarke (2011), p. 259.

[79] Asimov (1964), introduction; see also Weng et al. (2009).

[80] See, e.g., Weng et al. (2009); Anderson (2008), (2011)

[81] Clarke (2011), p. 260.

[82] Clarke (2011), p. 272.

go so far when it comes to the actual application of those search results. Presently, no AI on the legal market (see more below) possess the ability to implement moral or ethical judgment into their legal work, primarily in legal research, as a human lawyer would—which might bring a sigh of relief to human lawyers around the globe.[83]

It is worthwhile to examine this aspect of the legal profession because it has a large impact. Arguably, in the legal profession it is only judges who ought to be neutral, unbiased and free of such impulses in their jobs—lawyers represent a side in the process, they are technically not barred from being racist and biased, although the thought is repulsive. In addition to this, we must remember that law is of a didactic character—it is designed to teach and more often than not has a moral value behind its content. It undoubtedly means that all lawyers exercise influence in shaping the legal landscape and thereby hold a creative role in ameliorating it. This comes as something that is inextricably connected to an amelioration of ethical codes of conduct.

So, in this case, we could argue that an AI lawyer does not need to be completely free of bias either, but it instead would have to be able to distinguish when such behavior can impede them from making the correct decision and adhering to ethical codes regulating the legal profession. Suddenly it seems easier to find out which general and fundamental ethical constrains to place on an AI lawyer so that they are ethical by default. This is why I think it is important to place significant effort in researching and developing an appropriate ethical framework in which AI systems can operate, be they in the legal domain or otherwise (e.g., it is not difficult to imagine the magnitude of ethical implications of AI machines in the healthcare system).

This is an area of great importance in our further advancement of AI and I believe we ought to solve it sooner rather than later. It would be extremely detrimental to society to release AI systems, especially decision-making ones, that could deepen discrimination in society, on all levels, thereby turning back the clock on the anti-discrimination efforts done worldwide.

3.4 *Fairness by Design*

If these machines have the capacity of being more intelligent than humans,[84] if these machines are given the task to make decisions based on data processed, or in our case, if they are tasked with providing legal advice and representation, we need them to have appropriate ethical sensitivity. By saying "appropriate ethical sensitivity" I mean that this sensitivity is one that is arguably even better than the human sensitivity, but that idea is a more idealistic one and one that may take many years to achieve. In the short term, it would be beneficial to identify the manner in which different situations and problems can be represented in an AI system so that it can apply ethical solutions to them.

---

[83] Nunez (2015), p. 204.

[84] This is not the case today, where no AI is actually intelligent. See Storrs Hall (2011), p. 512.

The main idea that I propose here is to employ "Fairness by Design"—a term inspired by data protection law and its Privacy by Design principles,[85] This would entail that any and all AI systems are equipped with appropriate ethical sensitivity in the effort of eliminating bias or discrimination on any grounds. This needs to be done from the very beginning of creating an AI system, thus making sure that no system, anywhere in the world, would make distinction between people based on any discriminatory grounds and that the vision of machine ethics would be left intact: no humans can be harmed by the use of AI. It is, however, of vital importance to remember that old institutes and principles (legal, ethical or otherwise) change overtime to adapt to societal changes and the premise of this chapter is not to impose a static approach to ethics—on the contrary, we must grab the opportunity that lies in the fact that the field of machine ethics is in a developing stage and we must truly focus on creating sustainable, appropriate ethical solutions in this space. At the very least, it is our responsibility to see this through.[86]

In the end, our lives will be as good as the ethical principles, constrictions and sensitivities we place on the machines that will eventually encompass a big portion of our life as we know it today.[87] We do not want this to result in an adverse effect on humanity, and I think it doesn't have to, either. With proper efforts being put in now, in the embryonic phase of creating these advanced AI agents, we have the opportunity to see knowledge, skill, and morality and ethics evolve in ways more beneficial than we initially had intended.[88] In this sense, we are reminded of Asimov's story "Evidence" that largely inspired this chapter and the vision of a fully autonomous AI lawyer. In trying to discern whether Mr. Byerley was a robot or not, an interesting quip comes up about the nature of robots versus the nature of humans. The question comes up dealing with what mental differences there are between robots and humans, and the following answer provides us an insight into the thinking about ethical constraints of robots: "[Robots and men are] World different. Robots are essentially decent."[89]

The premise we employ here follows suit—robots and all AI systems we develop must have a degree of ethical constraints installed in the very core of the systems in order to properly perform their tasks, especially if they are lawyers or healthcare providers, and to eliminate possibilities of discrimination. Therefore, if these efforts are made now, in the beginning of this revolution, results will be more beneficial in the long run. This is not to say that the proposal of starting this process now will result in a fool-proof construction of AI lawyers (doctors, soldiers, etc.), but the errors will be detected more easily and more early on, with a greater chance of rectification.

I say this while being fully aware of the fact that, indeed, the ethics I am talking about relate to something we are yet to fully develop and, in that sense, there is incompleteness in this approach that pertains to the further development of these AI agents. Lastly, another question inevitably rises from all of this: if our AI lawyers

[85] See Footnote 8.

[86] Mackworth (2011), p. 347.

[87] Storrs Hall (2011), p. 522.

[88] Storrs Hall (2011), p. 523.

[89] Asimov (1950), p. 195.

are so technologically advanced, intelligent, and follow ethical and moral rules to a better degree than humans, do they get to have rights of their own? If yes, which ones? We must, alas, put a pin in this question as we cannot elaborate on it in the scope of this chapter, but it is a relevant discussion to have—and it encourages the Kantian argument of the importance of humans treating other humans, as well as other species in a correct and fair way, for the betterment of humanity in the long run.[90]

The fact that these agents would consistently be applying elevated ethical rules has the potential of improving society as a whole—by essentially improving our own application of ethical rules.[91] This is rooted in the fact that AI does not have emotions of any kind that would hinder them from picking and choosing between ethical rules based on personal circumstances.

4 Imagining the "Inhuman" Lawyer

This chapter explores the way AI is currently being put to use in the legal profession all the while addressing the common conceptions of how this will impact the profession in the next few decades. Furthermore, the chapter goes on to imagine a future AI lawyer, fully capable of conducting legal work in an uninhibited manner. The reader may wonder why embark on this thought experiment at all—but the answer is staring us in the face: the law is an ever-growing, comprehensive body of legal code, proving evermore difficult to navigate, with judicial systems all over the world suffering a great deal from saturation and legal professionals still working in the more-or-less same manner as they have been for a very long time.[92]

Furthermore, with society's constant stride towards technological advances that are designed to enable and empower our lives, I see no reason for law to lag behind. In fact, one can argue that a refusal to adapt and innovate may carry a degree of danger to society as a whole, as well as to law, as one of the essential pillars of society. Adapt we must, and an evolution of the way we operate will benefit us and make justice accessible to all.[93]

---

[90] Anderson (2011), p. 295; Kant thought that human-to-human behavior can deteriorate if humans are taught to mistreat other species, therefore it is in the interest of humanity that each human treat other species in a fair way. See Anderson (2011), p. 294.

[91] See Anderson (2011).

[92] See Bibel (2004).

[93] See Bibel (2004); Susskind (2017).

4.1 How We Use AI in Law Right Now

While law has never enjoyed a reputation for adopting new technologies,[94] we have at least embarked on a journey of making AI an integral part of the legal profession. Researchers in this space have long been somewhat frustrated at the slow pace of development in AI and Law.[95] Still, most efforts in the AI and Law domain have been focused on using AI to better convey legal norms to the public, i.e., to raise awareness and level of understanding of the law with the general population, thereby raising the amount of those who need legal advice and knowledge of the legal norms but may not necessarily have the financial means to afford such advice in the system as we know it today.[96]

This is an important step because it attempts to solve the conundrum of translating law into an algorithm. Furthermore, in the realm of what aspects of the legal domain (and profession) can be translated, it has been rightly observed that the work of lawyers is largely information processing which can be automated to a certain extent[97] which makes automation even more appealing if we look at it from the perspective of doing so with the aim to achieve greater efficiency and better coverage of the subject matter in each case. At this stage, what that means is an automation of more tedious work such as document review and information retrieval.[98]

So far, progress on creating more advanced tools in AI and Law has gone fairly slowly, especially compared to other areas where AI has been used more ubiquitously (e.g., virtual assistants like Siri and Alexa, bots, etc.). It has been suggested that there is a plethora of reasons behind this slow pace of development.[99] Namely, the sheer complexity of translating law and the different roles in the legal profession, as well as its concepts, rules, theories, judgments and interpretations into a capable system that can replace lawyers; the traditional way that law firms and lawyers bill for their services or the fact that, as is reiterated throughout this text, lawyers must be part of this development process from beginning to end and then be an essential part of the process of the actual work of these systems, including their upkeep and updates.[100]

Legal startups keep appearing with novelties in the way we work with cases, case law, discoveries and contract review.[101] Take ROSS for example: ROSS is the present version of an AI lawyer specialized in Bankruptcy Law, Intellectual Property and Labor Law. You ask ROSS a question and within a day, it will come back with a legal memo: ROSS (the system) does the research and creates a rough draft of a memo which is later reviewed and edited by humans.[102] ROSS is estimated to save

[94] Ambrogi (2017).

[95] See Oskamp and Lauritsen (2002).

[96] See also Leith (1988), p. 34.

[97] Remus and Levy (2016), p. 9.

[98] Henket (2003), p. 131.

[99] Oskamp and Lauritsen (2002), p. 232.

[100] See Footnote 99.

[101] Winick (2017); see also Cellan-Jones (2017).

[102] ROSS. Available at: https://rossintelligence.com. Accessed 27 April 2018.

twenty to thirty hours per case—a significant change in the way traditional law firms operate.[103] Created in less than a year,[104] ROSS represents a promising step forward into the possibility of creating fully independent AI lawyers that we are discussing in this chapter.

At the time of writing, ROSS seems to be the most advanced AI lawyer there is globally, but the truth remains that most AI systems used in law are document management systems of one sort or another.[105] Now, what ROSS and other advisory systems have in common is the fact that they do not make any decisions on the matters they research and analyze; instead, they focus on the analysis of factual circumstances that they are fed (by humans), they comb through relevant law and case law in order to bring an opinion forward—most often in the form of advice backed up by the analysis of relevant law, case law and the facts of the case.[106] What this essentially entails is that we have an opportunity to create really good legal assistants in any area of law, and these assistants would significantly cut down the time we need to comb through vast numbers of relevant text and combine the relevant pieces. This notion seems to cause two kinds of reactions with lawyers: one where there is genuine excitement over the time-saving nature of what this exercise entails, and the other where thoughts of unemployment spring to mind.

I am of the opinion that this is a welcome disruption in the legal profession and one that will not bring about unemployment but rather a redefinition of job descriptions. One often tends to forget that these systems will need human supervision, or "training," if you will. Human lawyers will be the ones helping the development of these systems using their skills, expertise and understanding of legal concepts.[107] These systems will have to be fed with vast amounts of information for a very long time. Even on their fully-operational level, as Hanket observes, it will be human lawyers who will be tasked with transposing factual circumstances of a given case into the system so that it can do its job.[108] Furthermore, the law is anything but stagnant. As an organism that is very much alive, law changes and develops often, leading to the necessity of constant oversight of AI systems—again requiring some degree of human intervention.[109]

[103]Nunez (2017), p. 193.

[104]Nunez (2017), p. 194.

[105]Oskamp and Lauritsen (2002).

[106]See also Henket (2003).

[107]Henket (2003), p. 128; Henket beautifully describes and exemplifies ways in which humans will be essential to the development and upkeep of these systems, at least for the foreseeable future.

[108]Henket (2003), p. 128.

[109]Henket (2003), p. 129.

4.2 How We Might Use AI in Law in the Future

If we are to move away from the mundane use of AI in the legal profession (i.e., document review etc.), we must bring lawyers on board with the idea of letting AI step into our world.[110] As already suggested, notions of AI lawyers tend to produce tinfoil hats out of thin air and this is something that the legal profession should work on a bit more: understanding where AI can be of extremely good use and help address time constraints as well as staffing issues, and putting it to use where needed. Ultimately, it is about selling the idea of the benefits of having AI legal assistants (and eventually lawyers) in order to get more lawyers interested in helping create these AI systems. After all, as has been said above, it is human lawyers that will be those who create, maintain and develop these systems. In addition to that, we would have to make significant advances in the space of neural networks and their ability to mirror human brains, in order to receive more out of the inhuman lawyer.[111] This is no small feat.

To imagine an AI lawyer operate without any hindrance—be it technological or ethical or otherwise—is by and large to imagine that justice is more affordable and more accessible to all. Furthermore, the ubiquitous presence and usage of AI lawyers will undoubtedly reshape the legal profession and I would argue that the reform would be for the better, in most part. Ultimately this would mean stepping away from the traditional billing-by-the-hour model and the like, originating in newer business models that will reflect the newly found efficiency of legal work through smart use of AI.

In addition to that, it has been suggested that putting legal cases in the hands of an AI system capable of processing them will result in more fairness since that will exclude human bias and because of its ability of processing more material than a human can, and can do so in a significantly shorter time span. It is well known that access to justice is all the more compromised nowadays. If we take the example of data protection law, we immediately take note of the lack of awareness and knowledge of individuals with regards to whom they share their personal data with, as well as what their rights are in this domain. This is not a good situation for anyone—much less for the average Jane who may not be aware of her rights due to the complexity of the system, the opacity and maze-like nature of the rules.[112]

In this regard, I have two things to say: (i) I would like for that to be our future as humankind—a more fair, accessible justice system on a global scale. Lower legal fees and bringing sound legal advice to those who would not be able to afford it in today's world. (ii) I will reiterate what has been said before me: these systems will, by and large, still need humans to feed them the factual circumstances of any given

---

[110]Oskamp et al. (1995).

[111]Neural networks are described as "an artificial imitation of their biological model, the brains and nervous systems of humans and animals. The ability to learn and the use of parallelism during data processing are important characteristics." Wettig and Zehendner (2004).

[112]See also Bibel (2004).

case, which ultimately means that the system will only be as unbiased as the facts presented to it by a human being.[113]

As I mentioned above, the didactic character of law is going to prove of utmost importance in this sense. On a general level, the role of lawyers is precisely to refine, polish the law so that it improves and adapts to society; therefore, lawyers will always be essential in the way of instilling legal values and refining the legal culture—this is the case now and this will be the case in the future when we are living in a world where AI lawyers operate as well as human ones. However, this point of view can be seen as a good compromise between those who would rather see an inhuman lawyer due to cost efficiency and speed, and those who would not like to be left out of a job in their lifetime: this point of view bridges society's needs to have a more accessible justice system and legal professionals' needs to have a job to go to on a daily basis.

5 Splitting the Baby

In the magnitude of the challenges at hand when trying to imagine an inhuman lawyer, we can apply a test that perhaps prima facie speaks more to AI judges but has the potential to give us insight into the kind of thinking a fully autonomous AI legal professional would have to be able to employ in order to do the job of a lawyer to the fullest extent of that job description.

The story of King Solomon and the baby is a well-known one and often used when discussing manipulative rules in order to uncover the truth.[114] In summary: in a maternity dispute between two women, both alleging to be the mother of a baby, King Solomon was left to employ one last, and a rather extreme method to determine to whom the baby really belongs: by threatening to split the baby in half and give halves of the baby to the women. Upon that threat, one of the women insisted on giving the baby to the other woman in order to keep the baby alive and this was a tell-tale sign that she was the true mother of this child.[115] For the purposes of this test, we will assume that Solomon's judgment was the right choice in this matter, and we will not take into account all the ways in which this can be disputed based on several factors, such as the issue of conclusiveness and completeness of evidence, as well as the role of deduction and logical forms such as modus tollens.[116] Instead, we will assume that there are no open-ended questions in Solomon's judgment and take it at face value.

The irony of applying a biblical test to a futuristic idea is not lost on me. Solomon's judgment fits in this situation because it is a relatively simple case with fragile, if any, evidence and its result is not a pure result of a legal discussion, but instead holds

[113]Henket (2003), p. 136.

[114]Leeson (2017), p. 41.

[115]See Footnote 114.

[116]For a good analysis of the role of evidence with reference to Solomon's judgment, see LaRue (2004).

elements of inter alia logic, probability and Solomon's acumen and, famously, his wisdom.[117] The first question we can ask is whether this case would have developed and ultimately ended differently if the king before whom these women stood was not as "wise" as Solomon was? Would this case have ended the same way if Hammurabbi had been the judge, therefore would lex talionis have been employed?

These questions indicate that there was no universal way to get to the conclusion that Solomon did—and there was no universal way to employ the sword-method that he used, which ultimately means that his reasoning had barely any legal basis and his decision and method were based on a hunch, or emotional intelligence if you will. Would an AI judge be able to do this? Robots being creatures devoid of emotions by default, the straightforward answer to this would be a resounding No. Can a robot curate a collection of experiences that would help them achieve this reasoning? It seems unlikely. If an AI judge were presented with two women and one baby, just like Solomon was, we can assume the following:

The AI would be able to process the request of the women and the juridical problem of determining motherhood. The AI judge would be able to identify the cause, subject and core of the dispute at hand. This would be done by the robot's capability to grasp the concept of disputes and legal norms. We can also assume that the AI judge would be able to create a test for the parties in the dispute, to make sure to bring the dispute closer to a resolution. This capability would likely be based on a concept of decision-making and deduction that would be instilled in the AI judge from a design stage where they are equipped with weighing between different options and choosing the ones that relate to the case the most.

Furthermore, we can assume that based on the first two capabilities, the AI judge would be able to discern which party in the dispute is pleading a stronger case. This would also be based on its capability to cross-reference the parties' arguments with relevant rules, both legal and ethical. However, the judgment on the case as was presented to Solomon would require more than what the AI judge could perform, it seems. Solomon did not use a specific rule to determine who the mother is, he used his ability to see through people and his accumulated experience.[118]

The challenge with facing an AI judge with a case like this lies in the completeness of its system—this AI judge would have to be extremely technically advanced to exercise what is basically common sense in order to issue a judgment. This is not easy for humans either, one has to be properly trained to make a certain kind of decision—not only in law but in general.[119] Therefore, this may be one of the limitations of AI in the legal domain. Perhaps we will be able to create legal AI that will do legal research and legal representation even, but judgments and similar decisions seem to be out of our reach for the time being because it would entail that developers would have to create ways for human elements such as emotion and common sense to be implemented and merged with the more structured sets of rules.[120]

---

[117]LaRue (2004).

[118]See Footnote 117.

[119]See also Clarke (2011), p. 280.

[120]See Footnote 119.

6 Conclusion

This chapter delved into a kind of daydream about the future of the legal profession. I do understand, however, that some may view this as more of a nightmare scenario, and I do hope that this widely-held opinion will eventually fade and be replaced by a more optimistic view on AI lawyers. Imagined as a thought experiment, the chapter touches upon various disciplines in the effort to understand the potential challenges we will face when attempting to create a robot lawyer.

The first question that this chapter deals with was focused on the (in)ability of translating law into an algorithm. This question opens a Pandora's box of possibilities, dilemmas and questions to be answered. Among them, linguistic aspects of such conversions were considered and as of yet, we have not succeeded in mirroring the logic, language and fluidity of the law into code that can help us transform the legal profession. At best, we have document processing systems, information retrieval systems and research systems such as ROSS.

In the pursuit of the aim of creating a fully functional AI lawyer, challenges of various nature will be met, such as: finding lawyers eager to co-operate in the development process, finding law firms and practices eager to try it out, pushing boundaries of natural language processing as well as IT in general in order to be able to technically convert this idea into reality, convincing lawyers and the public to actually use these AI lawyers, and lastly, ensuring a level of machine ethics and morality that we can deem satisfactory in order to allow these inhuman lawyers to advise us on legal matters and represent us in legal cases. With this in mind, how does the future of law look like?

One must always bear in mind that creating AI lawyers will be a long and difficult process, if creating them is at all a possibility. The challenges represented in this chapter are immense and, as of yet, we have no recipe for doing it right, thus it is bound to be a process of trial-and-error and one of baby steps, too; first, we will perfect the use of our AI legal research and information retrieval systems, expanding the legal domains in which they can operate so that they can provide substantial help to legal professionals in their daily work and cut down on time necessary to complete a task while, at the same time, raising the level of the service provided by virtue of the depth to which such systems can sift through relevant texts and come back with results.

This will, in and of itself, disrupt the legal profession and potentially bring us closer to being able to provide legal services to a larger demographic, not just to those who can afford top-notch legal advice. This phase will be characterized by advances in natural language processing that will aid our goal of creating a fully-capable AI lawyer eventually. In this phase, we will need to ensure proper translation and processing techniques. Human lawyers will still be heavily involved in the process, not only when it comes to the development of these systems, but in helping them operate, too. So far, so good. This will continue until we reach a critical point that will allow us to move on into another phase of building an AI lawyer.

The next phase will be the one of machine ethics. An examination of the necessity of installing ethics and morals into AI lawyers (or judges for that matter) will have to be executed and we will arguably fail in the attempt of creating a perfectly ethical AI lawyer at least a few times before we get closer to success. This is due to the fact that humans will, again, be heavily involved in this phase as well and that may result in flawed ethical principles and rules being installed in the AI lawyer. I propose a concept called Fairness by Design—meaning that a set of appropriate ethical standards will be deployed in any AI system, both in the legal domain and outside of it. This will be the minimum necessary to uphold the aim of machine ethics—after all, AI is not to impact humans adversely. However, the question does remain on whether we actually need perfectly ethical AI lawyers? In the sense of an AI judge, of course, bias will never be welcome and such systems should not be put in place, but for a regular lawyer—is a high ethical standard necessary or is it just an idealistic scenario, a nice-to-have? This can still be solved by using the Fairness by Design concept and further developing it to fit AI agents in the legal domain, thereby establishing the necessary number of ethical rules for them to abide by.

The future of law is not bleak—lawyers and legal professionals will still be heavily involved in the creation, maintenance and amelioration of AI in the legal domain, which means that the entire discipline has plenty of time to adapt and adjust to the reality in which fully autonomous AI lawyers operate. Even then, this may just be a job description redefinition, rather than an extinction of human lawyers. This is even more true of AI judges—based on the Solomon test, we have seen the numerous challenges for adjudication by an AI agent, some of which are heavily based on the ability of humans to employ common sense and contextualize situations in a better way than what we can imagine AI to be able to do, at least for now.

In sum, if the whole world is embracing AI, I see no reason why law should lag behind. This evolution of the legal domain, and consequently of the legal profession may indeed prove to be valuable for society in general and help close the gap of access to justice and other injustices currently plaguing the world.

Acknowledgements My gratitude goes out to Fredrik Persson and Andrej Bracanović for their input and immense patience during my many brainstorming sessions about robot lawyers.

References

Ambrogi, R. (2017). *Fear not, lawyers, AI is not your enemy*. https://abovethelaw.com/2017/10/fear-not-lawyers-ai-is-not-your-enemy/?rf=1. Accessed December 1, 2017.

Anderson, S. L. (2008). Asimov's "Three laws of robotics" and machine metaethics. *AI & Society, 22*(4), 477–493.

Anderson, S. L., & Anderson, M. (2007). *The consequences for human beings of creating ethical robots*. Sine loco.

Anderson, S. L. (2011). The unacceptability of Asimov's three laws of robotics as a basis for machine ethics. In M. Anderson & S. L. Anderson (Eds.), *Machine ethics*. Cambridge: Cambridge University Press.

Asimov, I. (1950). *I, Robot*. London: Harper Voyager.

Asimov, I. (1964). *The rest of the robots*. New York: Collins.
Baude, W., & Sachs, S. E. (2017). The law of interpretation. *Harvard Law Review, 130*(4), 1082–1147.
Bibel, L. W. (2004). AI and the conquest of complexity in law. *Artificial Intelligence Law, 12*, 159–180.
Bouaziz, J., et al. (2018). How artificial Intelligence can improve our understanding of the genes associated with endometriosis: Natural language processing of the PubMed database. *Hindawi BioMed Research International, 2018*. Article ID 6217812.
Brożek, B., & Jakubiec, M. (2017). On the Legal responsibility of autonomous machines. *Artificial Intelligence and Law, 25*(3), 293–304.
Cavoukian, A. (2010). Privacy by design: The definitive workshop. *Identity in the Information Society, 3*, 247–251.
Cellan-Jones, R. (2017). *The robot-lawyers are here—And they're winning, BBC news technology*. http://www.bbc.com/news/technology-41829534. Accessed January 1, 2017.
Clarke, R. (2011). Asimov's laws of robotics. In M. Anderson & S. L. Anderson (Eds.), *Machine ethics*. Cambridge: Cambridge University Press.
Farivar, C. (2016). *Lawyers: New court software is so awful it's getting people wrongly arrested*. Ars Technica. https://arstechnica.com/tech-policy/2016/12/court-software-glitches-result-in-erroneous-arrests-defense-lawyers-say/. Accessed January 1, 2018.
Fingas, J. (2017). *Parking ticket chat bot now helps refugees claim asylum*. Engadget https://www.engadget.com/2017/03/06/parking-ticket-chat-bot-now-helps-refugees-claim-asylum/. Accessed January 1, 2018.
Guarini, M. (2013). Introduction: Machine ethics and the ethics of building intelligent machines. *Topoi, 32*, 213.
Henket, M. (2003). Great expectations: AI and law as an issue for legal semiotics. *International Journal for the Semiotics of Law, 16*(2), 123–138.
Hodel, R. E. (1995). *An introduction to mathematical logic*. Mineola, New York: Dover Publications Inc.
Husa, J. (2016). Translating legal language and comparative law. *International Journal for the Semiotics of Law, 30*(2), 261–272.
IBM Systems & Technology Group. (2011). *White paper: Watson—A system designed for answers, The future of workload optimized systems design*. http://www-03.ibm.com/innovation/us/engines/assets/9442_Watson_A_System_White_Paper_POW03061-USEN-00_Final_Feb10_11.pdf. Accessed 1 January 2018.
Kac, M., & Ulam, S. M. (1968). *Mathematics and logic*. Mineola, New York: Dover Publications Inc.
Kasperkevic, J. (2015). Google says sorry for racist auto-tag in photo app. *The Guardian* https://www.theguardian.com/technology/2015/jul/01/google-sorry-racist-auto-tag-photo-app. Accessed January 1, 2018.
Kennedy, R. (2017). Algorithms and the rule of law. *Legal Information Management, 17*(3), 170–172.
Lame, G. (2004). Using NLP techniques to identify legal ontology components: Concepts and relations. *Artificial Intelligence and Law, 12*(4), 379–396.
LaRue, L. H. (2004). Solomon's judgment: A short essay on proof. *Law, Probability and Risk, 3*(1), 13–31.
Leeson, P. T. (2017). *Split the baby, drink the poison, carry the hot iron, swear on the Bible, adapted from WTF?! An economic tour of the weird*. Stanford: Stanford University Press.
Leith, P. (1988). The application of AI to law. *AI & Society, 2*(1), 31–46.
Leith, P. (1991). *The computerized lawyer: A guide to the use of computers in the legal profession*. London: Springer.
Mackworth, A. K. (2011). Architectures and ethics for robots, constraint satisfaction as a unitary design framework. In M. Anderson & S. L. Anderson (Eds.), *Machine ethics*. Cambridge: Cambridge University Press.

McGinnis, J. O., & Wasick, S. (2015). Law's algorithm. *Florida Law Review, 66,* 991–1050.
Mommers, L., et al. (2009). Understanding the law: Improving legal knowledge dissemination by translating the formal sources of law. *Artificial Intelligence Law., 17*(1), 51–78.
Moore, M. E. (Ed.). (2010). *Philosophy of Mathematics: Selected writings of Charles S. Pierce.* Bloomington and Indianapolis: Indiana University Press.
Nunez, C. (2017). Artificial intelligence and legal ethics: Whether AI Lawyers can make ethical decisions. *Tulane Journal of Technology & Intellectual Property, 20,* 189–204.
Nyholm, S., & Smids, J. (2016). The ethics of accident-algorithms for self-driving cars: An applied trolley problem? *Journal of Ethic Theory and Moral Practice, 19*(5), 1275–1289.
Oskamp, A., & Lauritsen, M. (2002). AI in law practice? So far, not much. *Artificial Intelligence and Law, 10*(4), 227–236.
Oskamp, A., Tragter, M., & Groendijk, C. (1995). AI and law: What about the future? *Artificial Intelligence and Law, 3*(3), 209–215.
Remus, D., & Levy, F. (2016). *Can robots be lawyers? Computers, lawyers and the practice of law.* https://ssrn.com/abstract=2701092. Accessed January 1, 2018.
Shulman, C., Jonsson H., & Tarleton, N. (2009). Machine ethics and superintelligence. In C. Reynolds, & A. Cassinelli (Eds.), *AP-CAP 2009: The Fifth Asia-Pacific Computing and Philosophy Conference*, Oct 1–2, University of Tokyo, Japan, Proceedings, pp. 95–97.
Stolpe, A. (2010). Norm system revision: Theory and application. *Artificial Intelligence and Law, 18*(3), 247–283.
Storrs Hall, J. (2011). Ethics for self-improving machines. In M. Anderson & S. L. Anderson (Eds.), *Machine ethics*. Cambridge: Cambridge University Press.
Susskind, R. (2017). *Tomorrow's lawyers*. Oxford: Oxford University Press.
Trevor, J. M., Bench-Capon, T. J. M., & Dunne, P. E. (2006). Argumentation in AI and law: Editor's introduction. *Artificial Intelligence and Law, 13*(1), 1–8.
Wallach, W., & Allen, C. (2008). *Moral machines: Teaching robots right from wrong*. Oxford: Oxford University Press.
Weller, C. (2016). The world's first artificially intelligent lawyer was just hired at a law firm. http://www.businessinsider.com/the-worlds-first-artificially-intelligent-lawyer-gets-hired-2016-5?r=US&IR=T&IR=T. Accessed December 1, 2017.
Weng, Y. H., Chen, C. H., & Sun, C. T. (2009). Toward the human-robot co-existence society: On safety intelligence for next generation robots. *International Journal of Social Robotics, 1,* 267–282.
Wettig, S., & Zehendner, E. (2004). A legal analysis of human and electronic agents. *Artificial Intelligence and Law, 12,* 111–135.
Winick, E. (2017). Lawyer-bots are shaking up jobs. *MIT Technology Review*. https://www.technologyreview.com/s/609556/lawyer-bots-are-shaking-up-jobs/. Accessed January 1, 2017.

Index

A
Accelerator programs, 98
Access right(s), 10, 123, 133, 140, 143, 148
Aerial regulation, 106, 116, 119
Agency, 17, 49, 93, 117, 118, 169, 175, 221
AI and Law, 211, 213, 226
AI ecosystems, 96, 97, 99
AI judge, 212, 229, 230, 232
Algorithm, 11, 33, 72, 73, 83, 126, 129, 130, 140, 142, 147, 148, 155, 157–160, 163–167, 170–179, 192, 196, 197, 209, 210, 212, 217, 218, 226, 231
Algorithmic collusion, 10, 156–158, 175, 176, 179
Animal(s), 8, 15, 18, 21, 24, 27–29, 34–38, 40, 41, 64–66, 70–74
Aristotle, 27, 217
Artificial Intelligence (AI), 1, 2, 5–12, 15–20, 22, 23, 25, 28, 31–33, 35, 38–45, 47–53, 57–60, 62, 63, 72, 75, 81–87, 90, 94–97, 99–101, 126, 131, 143, 149, 151, 156, 159, 175, 183–195, 197, 198, 200–207, 209–217, 219, 220, 222–232
Automated decision-making, 7, 9, 123–128, 130, 132–134, 136–138, 140, 141, 144, 145, 147–149, 151, 196
Automated processing, 11, 124, 133, 134, 138, 144, 145, 147, 183, 185, 194, 197, 205
Autonomy, 18, 20, 27, 32, 34, 43, 50, 60, 62–65, 71, 135, 143, 210
Aviation, 9, 105, 106, 109, 116–120

B
Big Data, 10, 18, 51, 87, 124–126, 131, 144, 150, 155, 156, 168, 177, 179, 186, 188, 190–192, 197, 201
Biology, 5
Black-box testing, 177
Bots, 2, 19, 47, 58, 169, 183–195, 197, 198, 200–207, 210, 222, 226

C
Callon, Michel, 88
Causation, 66
Certification, 52, 75, 115, 118, 201, 202
Chimpanzee, 28, 29
Citizenship, 16
Civil liability, 8, 15, 57, 72, 74–76
Code(s) of conduct, 115, 200–202, 223
Confidentiality of communications, 109
Conscious parallelism, 159, 161, 171, 172
Consent, 21, 51, 69, 108, 109, 137, 138, 147–150, 170, 192, 193, 196
Co-regulatory techniques, 205, 206
Criminal law, 21, 22, 34, 35, 38, 41, 65, 109
Cyberlaw, 58, 61

D
Data collection/Collection of data, 106–108, 111, 112, 114, 115, 178
Data protection, 7, 10, 112–115, 118, 123–126, 131–137, 139, 141–145, 150, 151, 184, 187, 188, 190, 197, 199, 200, 202–205, 207, 211, 224, 228

M. Corrales et al. (eds.), *Robotics, AI and the Future of Law*, Perspectives in Law, Business and Innovation, https://doi.org/10.1007/978-981-13-2874-9

Data Protection by Design, 114, 199, 204, 205
Data Protection Impact Assessment (DPIA), 200, 202
Delivery/Deliveries, 7, 89, 107, 115, 120
Descartes, 27, 29
Discrimination, 9, 10, 21, 35, 123, 128, 129, 132, 135–137, 146, 147, 149, 150, 156, 167, 168, 174, 175, 223, 224
Disruptive technologies, 9, 88, 94
Drones, 2, 9, 32, 48, 105–121
Dynamic regulation, 8, 81, 82, 88

E
Electronic surveillance, 109
Embodiment, 20, 60–62, 64, 65, 71
Emergence, 1, 61–64
Ethics, 7, 18, 25, 51, 210, 219–222, 224, 232
European Aviation Safety Agency (EASA), 117
European Group on Tort Law (EGTL), 66
European Union (EU), 7, 11, 42, 51, 75, 89, 96, 107, 109–112, 117–121, 169, 170, 174, 190, 195, 196, 200–202, 204
Experimentation, 89, 92, 93, 156
Explanation, 7, 47, 124, 130, 138–142, 196, 197
Extra-contractual liability, 8, 57, 58, 65–67, 76

F
Feld, Brad, 95
Financial Conduct Authority (FCA), 89

G
General Data Protection Regulation (GDPR), 7, 10, 11, 51, 113, 114, 121, 123, 132–150, 179, 183–185, 188–191, 193–195, 197–204, 206, 207
General legislation, 142, 197, 203
Geo-fencing, 115, 116
Goals, 38, 73–75, 164, 188, 203

H
Hawking, Steven, 16
Human control, 46

I
Incubator programs, 98, 100
Information obligations, 10, 123, 133, 138, 140, 143, 144, 148
Innovation, 1, 6, 8, 9, 16, 18, 58, 74, 81–83, 87–91, 94–99, 101, 119, 120, 166
Innovation ecosystems, 9, 81, 82, 87, 97, 98, 102
Insurance premiums, 127, 137
Intellectual property, 142, 148, 226
Interpretation, 43, 64, 125, 135, 147, 160, 169, 191, 194–196, 198, 205, 212, 215–219, 222
Investment, 8, 9, 81–84, 90, 92–97, 100

K
Killer robots, 16, 47

L
Law, 1, 4, 7, 8, 10–12, 15–26, 28–30, 32–34, 36, 38, 40, 41, 43, 46, 47, 49–52, 57, 58, 61, 63–75, 88, 89, 99, 108–110, 117–119, 132, 136, 138, 141, 143, 146, 148–151, 155–161, 165, 169–173, 176, 179, 183, 184, 187, 188, 190, 192, 196, 198–207, 209–219, 221–232
Law enforcement, 107, 108, 110, 111, 113, 115
Legal domain, 7, 11, 209, 211, 215, 223, 226, 230, 232
Legal entity, 24, 29, 30, 44, 48
Legal object, 26, 33, 44, 46, 48
Legal personhood, 8, 15, 19–22, 26, 31, 34–36, 38, 39, 41–43, 47, 49, 50, 52, 53, 67, 70, 74
Legal profession, 7, 11, 209, 211, 212, 222, 223, 225–228, 231, 232
Legal semiotics, 210
Legal subject, 22, 23, 34, 43, 44
Liability, 8, 15, 21, 24, 34, 39, 43–46, 48, 49, 57, 58, 61–63, 65–76, 117–119, 169

M
Machine ethics, 7, 219–222, 224, 231, 232
Minors, 21, 25, 33, 36, 49, 66, 68, 70, 71, 75
Misuse, 107–109
Monitoring algorithm, 157, 159
Musk, Elon, 16, 17, 33, 59, 97

N
Nature of law, 212, 218
No-fly zone, 116, 117

O
Objectification, 130
Opacity, 228

P
Parallel algorithm, 158
Per se illegal, 160, 176
Personal data, 9, 10, 51, 64, 108–110, 113–115, 123–125, 131, 132, 137, 138, 140, 143–145, 148, 150, 183, 184,

187–192, 195, 197, 198, 201–203, 206, 207, 228
Personalized pricing, 128, 168, 169, 174
Persons under supervision, 68
Policy experimentation, 88
Price fixing, 159, 165, 175
Principles on European Tort Law (PETL), 66
Privacy, 9, 10, 18, 24, 59, 105, 106, 108, 109, 111–115, 118–120, 123, 124, 137, 149, 150, 155, 178–180, 199, 202, 206, 211, 224
Privacy Impact Assessment (PIA), 113
Profiling, 9, 123–137, 140, 141, 143–145, 147–151, 185, 194, 197, 198, 204
Putin, Vladimir, 16

Q
Q-Learning, 163, 164

R
Recruiting, 128, 129, 133
Registration, 115–117
Regulatory methods, 202
Regulatory sandboxes, 8, 81, 82, 87, 89, 90, 101
Reinforcement learning, 163–165, 167, 177
Replicants, 71
Responsive regulation, 88, 101
Robot law, 8, 20
Robot lawyer, 216, 231
Robotics, 1–7, 11, 12, 16–20, 25, 31, 40, 46, 47, 51, 52, 58, 61, 63, 64, 97, 209, 210, 219, 221, 222
Robot(s), 2–5, 7, 8, 10, 11, 15–20, 22, 23, 26, 27, 29, 31–37, 39–44, 46–53, 57, 58, 60–76, 186, 209, 212, 220, 221, 224, 230
Roman slaves, 69
Rule of reason, 10, 155, 156, 172, 173, 176, 180

S
Science fiction, 33, 108, 186
Self-learning algorithms, 43, 157, 159
Signaling algorithm, 158
Silicon Valley, 85, 94, 95, 97, 101
Singularity, 16, 32, 83, 87
Social valence, 61, 63–65, 71
Sofia, 16
Soft law, 175
Specific legislation, 201, 203
Standards, 3, 16, 26, 50, 66, 115, 116, 120, 175, 176, 189, 191, 202, 203, 219, 232
Startups, 9, 81, 82, 85, 87, 95, 99–101, 226
Subject, 10, 16, 17, 20, 21, 24, 26, 28, 35, 39, 51, 63, 70, 71, 108, 113, 114, 118, 123, 132, 133, 136–145, 147, 148, 150, 151, 163, 184, 185, 190–201, 203–205, 212, 213, 215, 220, 226, 230
Surveillance, 4, 9, 32, 46, 105–112, 114, 115, 118–120

T
Tacit collusion, 10, 158, 159, 161, 164–166, 169, 179
Torts, 66
Trade secrets, 115, 142, 148
Transparency, 10, 18, 51, 110, 112, 115, 130, 139, 142–144, 148, 149, 151, 155, 158, 159, 161, 162, 167, 173, 179, 180, 189, 192, 197, 199, 205

U
Unintended collection of data, 114
Unintended data collection, 114
Unmanned Aerial Vehicle (UAV), 4, 106

V
Venture Capital (VC), 85, 92–94, 96, 97, 101

W
White-box testing, 177
Wiretapping, 109, 110

MIX
Papier aus verantwortungsvollen Quellen
Paper from responsible sources
FSC® C105338

If you have any concerns about our products,
you can contact us on
ProductSafety@springernature.com

In case Publisher is established outside the EU,
the EU authorized representative is:
Springer Nature Customer Service Center GmbH
Europaplatz 3, 69115 Heidelberg, Germany

Printed by Libri Plureos GmbH
in Hamburg, Germany